BRADY

Fire Chief's Guide to Administration and Management

BRADY

Fire Chief's Guide to Administration and Management

Richard A. Marinucci

Fire Chief, City of Farmington Hills, MI

Upper Saddle River, New Jersey 07458

Library of Congress Cataloging-in-Publication Data
Marinucci, Richard A.
Fire chief's guide to administration and management / Richard A. Marinucci.
p. cm.
Includes index.
ISBN-13: 978-0-13-613110-6
ISBN-10: 0-13-613110-7
1. Fire departments—Administration. 2. Fire chiefs. I. Title.
TH9158.M37 2009
363.37068—dc22 2008043473

Publisher: Julie Levin Alexander
Publisher's Assistant: Regina Bruno
Executive Editor: Marlene McHugh Pratt
Senior Acquisitions Editor: Stephen Smith
Associate Editor: Monica Moosang
Editorial Assistant: Heather Luciano
Development Editor: Jenna Caputo, Triple SSS Press
Director of Marketing: Karen Allman
Executive Marketing Manager: Katrin Beacom
Marketing Specialist: Michael Sirinides
Marketing Assistant: Judy Noh
Managing Production Editor: Patrick Walsh
Production Liaison: Julie Li
Production Editor: Karen Fortgang, bookworks
Manufacturing Manager: Ilene Sanford
Manufacturing Buyer: Pat Brown
Creative Director: Jayne Conte
Cover Designer: Margaret Kenselaar
Cover Photo: George Hall—Woodfin Camp/Woodfin Camp
Composition: Aptara®, Inc.
Printing and Binding: Edwards Brothers
Cover Printer: Phoenix Color Corporation

Figures 2.1, 2.2, 2.3, 2.4, 3.2, 3.4, 3.5, 4.1, 4.4, 5.1, 5.2, 5.3, 5.4, 6.1, 6.2, 6.3, 6.4, 6.5, 7.1, 7.2, 7.3, 7.5, 8.1, 8.3, 8.4, 9.3, 9.5, 10.1, 11.1, 11.2, 11.3, 12.1, and 12.3 appear courtesy of Mike Kish, Farmington Hills Fire Department.

Pearson Education Ltd., London
Pearson Education Singapore, Pte. Ltd
Pearson Education Canada, Inc.
Pearson Education—Japan
Pearson Education Australia PTY, Limited
Pearson Education North Asia Ltd., Hong Kong
Pearson Educación de Mexico, S.A. de C.V.
Pearson Education Malaysia, Pte. Ltd.
Pearson Education Upper Saddle River, New Jersey

10 9 8 7 6 5 4 3 2 1
ISBN-13: 978-0-13-613110-6
ISBN-10 0-13-613110-7

To my wife, Linda, and children, Jeff, Jill, Jessica, Reagan, and Sammy and to the Farmington Hills Fire Department

Contents

Foreword

Most fire chiefs had no clue they would end up in that position when they entered the fire service. Unlike the route to becoming a firefighter, which involves learning basic skills, the pathway to fire chief—and the knowledge, skills, and abilities necessary to be successful—is not so well defined. Being a crack fire suppression officer has little to do with being a good fire chief, and that's what gets many of us in trouble. Being the organizational leader requires a lot of finesse in dealing with people, communicating big-picture ideas, and leading sometimes-reluctant followers into the Promised Land. In addition, being cool under fire, especially when a commissioner or other elected official is calling you on the carpet, is paramount.

I spend quite a bit of time working with fire chiefs who are struggling in the deep waters of leadership, bureaucracy, and interpersonal dynamics. This text has great tools and information to assist you in being a good fire chief. Successful chiefs understand that winners get what they want by helping others get what they need, or in other words, supporting the organization's needs is how you can be successful. Individuals who are self-centered and not focused on the groups they lead will ultimately fail.

So how does one get the response needed from an organization of free spirits, free thinkers, and overachievers? It is necessary to understand the psyche of the emergency responders and the other supporters of the fire department and build a system of support for them. Not only are you going to need those skills, but the others whom you lead will need them as well. If one of the assistant chiefs can't "play well with others," then your haunting has just begun. If new employees don't know what is expected of them, then the organization gets to "herd" them around for their 25-year careers. If the administrative assistant is the point of contact for department rumors and discontent, then you again have a "gift that keeps on giving." These are a few of the challenges I see continually in fire service organizations.

Chief Marinucci will give you the tools and knowledge for dealing with these situations effectively in this text. He is a great leader not only in his own department but at the national level as well. As the previous president of the International Association of Fire Chiefs he has traveled extensively in search of lessons learned from good leaders. As a former chief operating officer of the U.S. Fire Administration he has seen the challenges the fire service faces on a national and international level. I don't believe there is a more qualified person to write this text.

So, enjoy this gift to you and immerse yourself in the stories and information contained within. Thanks, Rich, for writing this and thanks to you, the reader, for deciding to improve the profession of fire chief.

Jeff Dyar
Fire Commissioner, Upper Pine River Fire Protection District, Bayfield, CO. Program Chair (ret.), EMS—Firefighter Health and Safety-Counterterrorism, National Fire Academy, Emmitsburg, MD

Preface

When the author became fire chief, the local newspaper headline proclaimed that he had become the "top firefighter." It sounded good. If only that were the job, it would be relatively simple: respond to a fire, extinguish it, and wait for the next one. As most fire chiefs have discovered early in their career, the job description for fire chief is much different from that for firefighter. Having a guide to the job would be very helpful. Not something that is too burdensome to read, just something to help and act as a simple reference for the issues that happen regularly. Although having a book with many examples to help with the process would be beneficial, the real need is to consider the "softer" skills required of the job. The job is about thinking, reacting, problem solving, and the interpersonal relationships that are critical to being successful. The title of this book suggests the content based on these premises.

Often, new chiefs can be overwhelmed when they reach the top of the profession, not realizing the types of things that will be expected. They usually find that researching the formal and official details of the job is not especially challenging compared with what they discover when the day-to-day responsibilities start to pile up. Technical documents and non–fire service publications are very helpful and important for finding examples and samples of the typical duties of a fire chief, yet they miss some of the key components of success. Besides problem solving and thinking skills, the successful chief needs to understand the importance of politics and personal relationships. The fire service is unique, so the chief executive or fire chief has slightly different needs than someone in management and administration in any other type of business or industry. The brotherhood (and sisterhood), work schedules, and the extensive use of volunteer firefighters present challenges not present in other professions.

Fire departments exist because someone has to respond when a person calls 911. If there were no emergencies, there would be no need for a fire department. Everything a department does supports that emergency response function. Equipment and apparatus, training, and even fire prevention activities exist only because a citizen needs help fixing a problem, usually relatively quickly. Even more important is the need for the department to be really good at what it does. No one asks for a mediocre response when they have an urgent problem. They expect an "A" performance. The fire chief needs to do what is necessary to prepare the organization to be outstanding all the time and on every call. Skills and tools needed to be successful can be acquired when they are lacking and enhanced when already present. Perfection is the goal. This means that the chief also must be good all the time and must continue to improve the department's performance. Good advice is, don't rest on your laurels and don't believe all your press clippings! Although nice to receive, the last compliment does not guarantee future kudos.

A tremendous amount of responsibility is placed on the fire chief, regardless of the size or type of organization. The chief must provide leadership to take the department to level of service expected in their community. The chief is not necessarily expected to ride on the trucks or provide service on the scene of an emergency—though in many communities that is part of the job—but to make sure the resources are available. This includes personnel and their preparation, apparatus, equipment, and various support functions. Fire departments are not islands (or shouldn't be) and rely on others outside the organization to help provide quality service. In addition to responding to fires, they provide emergency medical service (EMS), hazardous materials mitigation (hazmat), technical rescue, and disaster response. Clearly, the job is complicated and requires good skills to be efficient and effective.

Unfortunately, once promoted, the fire chief often does not have the time to obtain the necessary education for the job. Reference materials that are too lengthy, complex, and formal are not likely to be read from cover to cover. They take up space on a bookshelf and are opened only on special occasions. This simple text outlining the basics will be very beneficial toward providing the essentials needed to be successful from the very beginning.

Successful chiefs do not expect all the answers to be laid out in a simplified format, nor do they expect to find any source that answers every question or fixes all the problems they are likely to face. This text is intended to be a relatively easy read that touches on the necessities of the job. Other sources allow even more detailed study on specific topics, including other books specific to the fire service and generic to management and administration.

The expectations for performance of fire chiefs are higher than they have ever been in the past. The job is more complicated, requiring different skills than may be required on the scene of a working fire. Rarely are the firefighting capabilities of the chief questioned. Chiefs are challenged in the other areas of their responsibilities: human resource management, politics, ethics, labor relations, budget and finance management, and their ability to lead change to adapt to the very changing world. Fire chiefs need to think quickly and adapt on the run. They are expected to fix things when they go wrong. This text is designed to provide the basic tools to address these critical job functions of the modern fire chief. It is hoped you find help in gaining insight whether you are just getting started or have a few years under your belt.

Acknowledgments

To someone who isn't a writer, taking on a project like this becomes overwhelming at certain points in the process. There are many times when you question whether you should continue your efforts. Fortunately there are people in your life who offer support, guidance, encouragement, and a smile from time to time. I thank my wife, Linda, who I am sure doesn't realize the contribution she made. I also appreciate everyone at the Farmington Hills Fire Department. The 24 years spent there as fire chief provided me many opportunities and experiences, which are the basis for this book. Special thanks to Jean Coil for her help and to Inspector Mike Kish, who took my photos and found examples in the department files to use. Lastly, I must acknowledge Brady Publishing and Katrin Beacom, Stephen Smith, and Jenna Caputo, who did their best to get me started and keep me on schedule. There are others, locally, nationally, and internationally who contributed in some way. There are too many to acknowledge here, but I hope they know who they are.

REVIEWERS

John P. Alexander, Adjunct Instructor
Connecticut Fire Academy
Windsor Locks, CT

Bernie Becker, MS, EFO, CFO, NREMT-P
Clearcreek Fire District, Chief of Department
Springboro, OH

Jeff Dean, Assistant Chief
GPSTC Fire Department
Section Supervisor, Georgia Fire Academy
Forsyth, GA

Mark Martin, Division Chief (Retired)
City of Stow Fire Department
Stow, OH

Gordon M. Sachs, Fire Chief (Retired)
Fairfield Fire & EMS
Fairfield, PA

Brad Van Ert
Captain / EMS Coordinator
Downey Fire Department
Downey, CA

David Williams, EMT-P
Commissioner (Past Chief Officer), WCFD4
Bellingham, WA
Safety Manger, VECA Electric and Communications
Seattle, WA

About the Author

Richard Marinucci has been the chief of the Farmington Hills Fire Department since 1984. As chief, he is responsible for the administration and management of the department, which includes Emergency Medical Service (EMS), fire prevention, public fire safety education, training, emergency management, hazardous materials, and technical rescue.

Chief Marinucci was president of the International Association of Fire Chiefs, a 12,000-member organization representing fire chiefs and chief fire officers, for the 1997–98 term. He was the first chair of the Commission on Chief Fire Officer Designation, currently known as the Commission on Professional Credentialing. He is currently designated as a Chief Fire Officer (CFO).

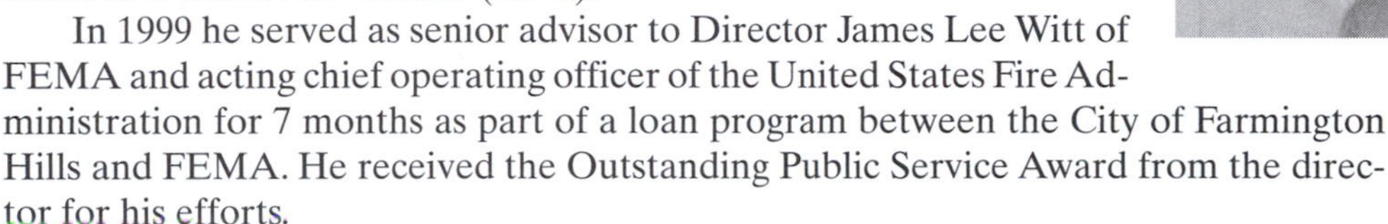

In 1999 he served as senior advisor to Director James Lee Witt of FEMA and acting chief operating officer of the United States Fire Administration for 7 months as part of a loan program between the City of Farmington Hills and FEMA. He received the Outstanding Public Service Award from the director for his efforts.

Chief Marinucci has three bachelor of science degrees: in secondary education from Western Michigan University, in fire science from Madonna College, and in fire administration from the University of Cincinnati. He was the first graduate of the Open Learning Fire Service Program at the University of Cincinnati (summa cum laude) and was named a Distinguished Alumnus in 1995. He has since coauthored a revision of two of the open learning courses. He has also attended courses at the National Fire Academy.

Chief Marinucci was selected to be a panel member of the Recommissioned America Burning committee.

He is a member of the Western Wayne County Fire Chiefs (president since 2005), the Regional Alliance for Firefighter Training (chair since its inception in 1997), Oakland County Fire Chiefs, Southeastern Michigan Fire Chiefs (president 1989–90), Michigan Fire Chiefs, and National Fire Protection Association. He served as president of the Great Lakes Division of the IAFC for the 1992–93 term. He was coeditor of the *Michigan Fire Service News* from 1989 to 1996. He currently is a regular contributing author for *Fire Engineering* and *Fire Apparatus* magazines.

Chief Marinucci has been an instructor in a wide variety of training programs from basic firefighter to chief officer training. He has been an adjunct faculty member of Madonna University, Eastern Michigan University, and Oakland Community College. He has presented programs for the International Association of Fire Chiefs, California Fire Instructors Workshop, Fire Department Instructors Conference, Fire Chiefs Association of Japan, the National Fire Academy EFO Symposium, and the Maryland Fire Rescue Institute National Staff and Command School, and numerous regional, state, and local conferences and workshops across the country.

BRADY

Fire Chief's Guide to Administration and Management

Introduction

What exactly does a fire chief do? If you looked at a typical calendar, you would think the job was to go to meetings! I am sure there are many thoughts on this, but the job descriptions and definitions are probably somewhat similar at their core. I am equally certain that what actually happens in the real world is different from the job description. If you ask a fire chief what he or she really does, you will get a laundry list of items—budgeting and purchasing, personnel issues and human relations, politics, customer service, labor relations, and maybe even planning. The list may be long or short, general or detailed, serious or tongue in cheek, but what you won't typically hear is "fight fire," at least not in the literal sense (see Figure 1.1). Being a fire chief requires skills in leadership, management, communication, and teamwork. Sometimes a chief will work alone; however, all chiefs need a core set of skills that allows them to do the basics of the job. Ultimately, they are expected to produce results that make the community more livable by minimizing the impact of fire and improving on the other first-responder services such as emergency medical service (EMS), emergency management, hazardous materials response and mitigation, and a myriad of special rescue situations.

The fact is that most successful chiefs have little direct involvement with on-scene emergencies. They are responsible for preparing the organization to respond to the endless stream of incidents and calls for service, as well as the nonemergency functions such as fire prevention activities. Simply put, the chief's job is to provide the resources required to provide the services expected. The resources include properly trained and prepared personnel, appropriate apparatus and equipment, and policies and procedures to guide the organization. An equally important responsibility of the job is to elevate the professions of firefighter and fire chief, in particular. Both need to evolve and to improve continually to meet the needs of the community served. Basically, citizens want perfect service when they have an emergency and call 911. They want the "A" team every time. Members of the community are no different from those in the fire service profession who would expect nothing short of excellence if one of their loved ones needed service. Fire chiefs must continue to raise the bar with respect to the quality of service provided locally and throughout the profession.

With this said, what has been done to prepare a chief to handle the responsibilities of the job? Often, it is very little or nothing. Individuals are frequently promoted for the work they have done, not for what they will do. Although past performance is usually an indicator of the future, it does not always guarantee success. Previous work

FIGURE 1.1 ◆ Five bugles signify the rank of chief.

history is essential in determining abilities; however, knowing the job at hand and preparing properly for it contribute greatly to success as a chief.

Certainly there are many reasons why some chiefs survive and even flourish while others barely survive. Those who do well understand that the job is about personnel issues, politics, labor relations, politics, acquiring and utilizing resources, politics, communications, and to some extent, luck. (I hope you noticed the multiple references to *politics*. There will be more on that subject later, but suffice it to say that politics is a critical component of the job, yet little formal preparation is offered.) The successful chief needs to be prepared with a combination of experience, education, and the establishment of a network to aid with the job. Thus, in addition to being able to deal with the characteristics and intangibles unique to the community a chief must possess the following general skills:

- Political
- Leadership
- Operational
- Financial
- Human resource
- Customer service
- Public relations
- Interpersonal
- Sales
- Promotion

As a chief you wield much influence, much attributed to the position, not necessarily to you as an individual. You build on (or in some cases destroy) that influence based on your performance. Bear in mind that you will be judged as much or more on *how* you do things than on what you actually do. You will constantly be making decisions—some critical, some not; some instantaneous, some deliberative; some clear-cut, some controversial. You need to develop patience when timing is required but be able to act quickly when circumstances require urgency. Much of the job can be very easy, but it is your actions when times are tough that will make or break you.

Of course, you must realize that as you ascend the ranks you actually lose some rights. Because of your position, there are certain things that you are expected to do and others that are not in your best interest. For example, you lose a little of your freedom of speech. Although the United States Constitution guarantees you certain rights, sometimes things are best left unsaid if you want to keep your position. Everything (potentially) that you say can be scrutinized, analyzed, and interpreted for the benefit of someone else (either friend or foe). The media, employees (union), bosses, politicians, or anyone who thinks that your opinion is important can use your words. They can be used to contribute to rumors or be taken out of context. This is not to prohibit you from speaking but to remind you to be more conscious of your position when you make a statement.

The job of chief of the fire department has changed greatly over the years and most likely will continue to evolve (see Figure 1.2). The world around the fire service is changing very rapidly, much faster than it used to. More external forces are being exerted on the fire service than ever before—citizen, taxpayer, and political demands; state and federal mandates, standards, and laws; liability issues; diversity concerns; and requests for service outside the traditional role of fire departments.

Fire chiefs are managers, so they must have management skills to provide the level of service desired by the policy makers in the community as well as what the citizens want to support. From this perspective, it is the job of the fire chief to carry out the policies established by others, such as staffing levels and budgets.

FIGURE 1.2 ◆ The chief has much more to do than just be the top firefighter.

In addition to being a good manager, a chief needs to be a strong leader. Although leadership is not necessarily important in performing as the top manager of the organization, it is required to influence direction and policy. The chief's leadership skills are important in getting the entire fire department team to pursue excellence and will determine the level of success attained by the department. Although a chief can survive as one or the other or have limited abilities in either area, the ideal chief will be both a great manager and an outstanding leader and will succeed in meeting high expectations by using these skills appropriately.

Sometimes, though, it is necessary for the chief to be a follower. Although he or she may be the "top dog" in the fire department, the chief is most likely a member of the middle management of the larger governmental agency of the community. Just as many skills are needed to be a follower as to be a leader, and it is important to recognize which role is appropriate at any given time. Middle managing is difficult, with pressures from both ends. A chief must be supportive of the department while being part of the management team.

In addition to the skills already described, more personal attributes—the ability to handle pressure and stress, take charge, make decisions, and remain optimistic and enthusiastic about the job of chief—are critical to success. Although most of the skills can be learned and the knowledge attained, ability is also necessary. Just as it takes more than being tall or practicing incessantly to be a good basketball player, so, too, does becoming a successful chief mean recognizing and developing the intangibles associated with the job. It helps to remember the purpose of the chief position—to provide service and leadership—and that the role is voluntary. Choosing the job means accepting the responsibilities that come with it.

Simply put, a successful chief understands management, leadership, and communications, and is skillful in all. There is no specific chapter devoted to any of these topics, though each is a critical component of everything related to the job. The reason for this is twofold. First, there are many books, texts, and resources such as courses, workshops, and seminars committed to these topics. There is no lack of information on these topics. The fire chief needs to recognize the value of learning as much as possible about these subjects and to pursue them not only in preparing for the job but continuing until retirement. The basic principles of good leadership, management, and communications are valuable in all careers. Second, the topics herein are designed to support leadership, management, and communications skills. There are times to lead, times to manage, and times to do both. Communications are important all the time. As you work through the text, keep in mind the need for all these skills.

Although leadership is not a separate topic or chapter, there are some basic things you can do and consider as you move through this text. First, recognize that there are many styles and personalities of leadership. Great leaders have their own way of doing things. You will need to maintain the core competencies of leadership but demonstrate them through your personality. For example, Generals Eisenhower and Patton were both great leaders within their chosen professions, yet both had distinct personalities. You cannot copy styles if they do not fit your makeup. The second thing you can do to work on your leadership abilities is to learn from leaders close to you. As you were promoted through the ranks of the fire department, who provided a good example for you to follow? What skills did they possess? Whom else do you observe in positions of leadership? Take stock of how they fulfill their responsibilities as a leader. They know their job and continue to improve. Take the time to list the traits of the leaders whom you respect and work toward developing your own abilities accordingly.

Managers are important in all organizations. You cannot delegate your management responsibilities. Likewise, you cannot excel as a chief without being a great leader. As the leader, you are entrusted with great responsibility. Accept the privilege. You will meet people along the way whom you believe to be outstanding chiefs. Others will make you wonder how they got the job. Those who accept the responsibility and privilege will strive to be great. You will need to understand your role and also to work to continually improve yourself and your organization.

Being a chief is an art. There are few black-and-white issues and a whole sea of gray. All answers are not found in a textbook, solutions do not fit all problems, and no two chief positions are identical. Communities and fire departments differ and require individuals who are creative and adaptive to lead their fire departments. If it were easy, anyone could do it. Prepare yourself for the position and always strive to grow and improve. Shoot for perfection, which will encourage you to improve every day.

The purpose of this book is to get you to analyze the job, to get you to think, maybe even to inspire you, and to offer you some tools for competing in the real world. This does not mean that the answers are going to be provided. Those are for you to discover. Successful chiefs do not always apply standard solutions. They know when to think outside the box and when to go with the flow. A chief needs the ability to think, reason, analyze, take action, have patience, solve problems, find resources, and build a network. If individuals are aware of the expectations of a job, they can better prepare.

As you proceed through this book, think of how you might address circumstances instead of looking for ready-made solutions. A chief does not need to look in this book for a job description. He or she needs to know why one is needed and what might be the best source for his or her department and community. Finding examples is easy. You need to know which one is most applicable to your organization. In addition, discovering information on your own will be more valuable in your overall preparation for the job. There is no shortcut, but good instincts, experience, and education will lead to success.

Politics

KEY TERMS

politics, p. 5
partisan politics, p. 5
political relationships, p. 11
politicians, p. 6

Politics is arguably the most important part of a fire chief's job. Chiefs are primarily (in most cases) administrative personnel who must exercise political prowess and who will be required to be politicians more than they might like. There is an element of politics in every job in the fire service and it becomes more significant, more important, and more time consuming as you ascend the ranks. By *politics* is not meant **partisan politics**, which involves strong allegiance to a particular political party, but the core meaning of the term. *Webster's Collegiate Dictionary* (Tenth Edition) offers a few definitions of politics that relate to its importance in the role of the fire chief:

- The art or science of government
- The art or science concerned with guiding or influencing government policy
- Competition between competing interest groups or individuals for power and leadership
- The total complex of relations among people living in society

politics
The art or science of government; the art or science concerned with guiding or influencing government policy; competition between competing interest groups or individuals for power and leadership; the total complex of relations among people living in society

partisan politics
Strong allegiance to a particular political party

You can see from these definitions that a fire chief needs to develop skills in this art or science of government to be effective. If you examine these four definitions, you see some common elements and key words. First, politics is an art (though science is mentioned, in the arena of fire it is definitely an art form). Next, it involves influencing government policy. (Perhaps members of the fire service should pay more attention to this part of the job to get the resources necessary to do the job.) The definitions also include the words *power* and *leadership,* qualities already discussed as being important to the successful fire chief, and, lastly, *relations.*

Not mentioned in the definitions is the concept of right or wrong. As a professional, you often see the end result of the process, but the political process is just as important as the final outcome. Often, the "right" thing is identified, but the system does not allow it. Decisions can be right, wrong, or political. For many people this is a difficult concept to grasp. To those not in the political arena, the answers seem so obvious and decisions so simple; but in the world of politics, things are never as they seem. The web is so complex that actions create reactions in many other, often seemingly unrelated, places.

Combined, the definitions of politics offer a reasonable job description for the position of chief—build relationships in a complex organization, guide and influence policy, and provide leadership. There may be some debate among fire service personnel as to the importance of politics, but if you accept the fact that leading people and obtaining resources to do the job is the role of the fire chief, then politics needs to be at the top of the list.

THE BASICS

Most fire chiefs are promoted to the position without adequate preparation. If you were to ask several current chiefs how much of their job involves politics, they would tell you anywhere from 50 percent to 100 percent. Comparing that level of involvement and the importance of politics to the job with the amount of education and training the chiefs have received reveals a large disconnect. If chiefs are successful, they have a natural affinity for the role and have good instincts regarding the combined definition of politics, but there are still things to be learned. As in so many other areas, recognizing the importance of politics as a major component of the fire chief's job is the first step toward preparation. Another key point is acknowledging differences in politics from one community to another depending on the history and culture of the organizations that make up the community. Among the commonalities shared by successful fire chiefs who recognize, understand, and accept their role with respect to politics are the following:

- Acknowledgment that politics is a key component of the job
- Self-assessment of personal skills, knowledge, and instincts
- Education, both current and future
- Experience
- Network development in and out of the fire service
- Awareness of priorities in "choosing battles"
- Acceptance that you don't need to win, even when you can

politician A person formally engaged in politics

Many in the fire service shy away from politics for a variety of reasons. First, the fire service is a very trusted agency, whereas a **politician**, a person formally engaged in politics (and to some extent general government) may not be respected or trusted by the general population (as indicated by additional definitions listed in the dictionary). Many citizens are so disenchanted with politics that the mere mention of the word conjures up negative thoughts. Why would members of the fire service want to be part of something so unpopular? The word *politics* also has a very negative connotation within work groups. Asking why various things are wrong in an organization is often answered with a derisive "politics" by people within (or outside) the organization. Why did one person get promoted over another? Those who didn't agree with the decision again answer "politics."

Another reason for the lack of attention to politics is the failure to recognize its importance and to prepare adequately. You cannot be comfortable doing this part of the job unless you have sufficient skills to be effective and have the confidence to participate. Further, you can be effective in politics and still remain respected and maintain your image (and that of your department).

What is involved with politics? There are two general components of politics, formal and informal. Formal politics involve the elected and appointed officials at all levels of government: federal, state, county, and local. Informal politics involve the relationships with all the individuals and organizations who interact with the fire department.

◆ POLITICAL STRUCTURES

There are many books on the organization and structure of states and the federal government. You may even recall some things from your high school civics class. All levels of government are organized essentially the same, with an executive head and a legislative branch. The federal government has the president as the chief executive to carry out policy and the Congress to legislate. States have governors and some legislative structure—full or part-time, bicameral, and so on. Cities have either a strong mayoral form of government or a council–manager system (which has various hybrids). In cities with a strong mayor, there is a legislative body—the council—that theoretically establishes policy for the chief executive, in this case the mayor. In a council–manager system, an elected body not only sets policies but hires a professional manager to carry them out. There are other variations of government. For example, in a township system, a board of trustees and a township supervisor may carry out the legislative and executive functions. You need to know which system is used in your community and the specifics of its organizational structure. This is important for many reasons but especially to know who has the responsibility for setting policy and appropriating funds.

◆ FORMAL POLITICS

Formal politics is closer to the definition "the art of government." The fire chief needs to know the political structure of government on the local level, as it establishes the desired local power structure. It also establishes the method for adopting ordinances, codes, and local laws. The local government charter most likely also establishes the rules for the organization of the fire department (and other city services such as police). Although there is an established power structure in political systems, fire chiefs can play in the arena and can have influence if they understand the system and the players. What is your local form of government? Who are the power brokers? Who has access to the formal political players and who has influence? See Figure 2.1.

As fire chief you need to understand what role politics plays in establishing policy and budget. It is then your job to carry out that policy. You have an opportunity to influence policy, but in the end your job is to implement it. Accepting this fact will help not only in maintaining relationships and job stature but also your mental health and stress factors. Fire chiefs can get very frustrated when things don't go the way they think they should. Professionally they may think that they do not have the tools to provide the level of service needed in the community. In politics it is important to remember that you do not need to win, even if you can. Regardless, if you are unable to influence the policy makers, then you must accept their direction. If you cannot or will not accept their direction, dust off your résumé!

There are many caution signs that you as the fire chief must be aware of. Be careful with whom you align. Politics can be fickle, and politicians serve at the whim of the electorate. Direct contact and affiliations with particular parties or individuals can

FIGURE 2.1 ◆ Elected officials at all levels of government establish policies that affect fire departments and fire chiefs.

affect your future when there is a changing of the guard. Developing apolitical power is the ultimate goal of a chief.

In addition to local politics, other layers of government have a role in the fire and emergency services provided. Although they may not directly provide funding or establish your right to exist as a fire department, they affect your operation. Perhaps they make grant funds available or provide money to your local government for other services. This could free up resources for the emergency services. There are also many acts, laws, and administrative rules that apply to the fire service. Think about the numerous federal and state directives you must comply with—OSHA rules; licensures; EMS regulations; antiterrorism requirements to access funding, training standards, and requirements; fire safety codes; and a host of other lesser known issues. Often, these directives are the result of the actions of a few, yet they affect everyone. The point here is not to debate the merits of the various laws and regulations but to acknowledge their existence and impact on your job.

Failure to understand laws, rules, policy, and standards can land you in trouble. Like it or not, you must follow the law. Further, failure to abide by established standards can create a legal liability for you, personally and professionally. However, you can have an influence on a county, state, and national level. Although it is more challenging to be involved in the political process as you get farther from local government, you cannot ignore it, as policies, laws, rules, and regulations affect your ability to provide service to your community. The same principles that apply locally are the same throughout the various political systems.

BECOMING INVOLVED IN FORMAL POLITICS

How do you get engaged in the formal politics? First, you must understand the reasons people desire to seek election or work in formal political positions. Although some people are truly committed to public service, others choose to enter the political arena for power, ego, influence, and money. Knowing this helps in establishing a strategy to work within the system.

One of the primary goals of elected officials is to get re-elected. How do you and your department deliver votes? Some do it by campaigning. Some contribute financially. Some work behind the scenes for their candidate.

The most common fire service involvement in formal politics is generally by employee groups or labor unions. In states where there is no collective bargaining and/or labor unions are not recognized, employee groups such as firefighters may still elect to become engaged in the political process in their community or state. Certain volunteer fire agencies, because of their history and culture, traditionally remain engaged in local politics. Sometimes the fire chief or top fire official (e.g., commissioner) is active in formal politics, especially when the fire chief is appointed directly by an elected mayor. These circumstances often require the fire chief to be part of the mayor's campaign, participate in fundraisers, and make donations.

Effective use of the political system by such groups will definitely have an influence on the operation of the department. Depending on the strength of the political alliance, there may be the perception that someone other than the fire chief actually runs the fire department. Successes in the formal political arena allow access to politicians. This access is important in influencing policy. Ultimately, the fire chief must carry out the policy established by the political body. Groups influencing policy gain personally and professionally. In many instances, the fire chief cannot play by the same rules, either by law or for self-preservation. Backing the wrong candidate is not good for job tenure!

If you still have the desire to engage in the more formal process of politics, research your state political contribution laws and regulations. Although you have certain constitutional rights, you may specifically be prohibited from using any public resource for campaign purposes. For example, if you were to drive a department-owned vehicle to a fund-raiser held at a fire station community meeting room, you might be asking for trouble! The chief must accept the circumstances and understand what can and cannot be done (or even what should or shouldn't be done). Failure to recognize the influence of employee groups in the community has led to the downfall of many fire chiefs.

There has been some erosion of political clout owing to the elimination of residency requirements for firefighters. Through negotiations firefighters have earned the right to live where they choose, but this benefit has the potential to reduce the political clout of the firefighters. In contrast, most volunteer fire departments maintain their residency requirements, for obvious reasons. Thus a strong bond is maintained in those communities between the fire department and the political body.

What actions can be taken to counteract the unintended consequence of loss of political clout? Some fire departments do it by providing quality essential service for the health and welfare of the citizens (voters) that becomes indispensable in the community. This generates the grassroots support that can be vital to the department. It also gives the elected officials something about which to brag. This is not to say that the fire and emergency services are not intrinsically important—not only to the community but to the politicians (see Figure 2.2).

FIGURE 2.2 ◆ Fire departments can remain popular with good service not only at emergencies but through scheduled and planned events.

The popularity of the fire service, especially since the events of September 11, 2001, would seem to strengthen its political position. It has become popular to associate with firefighters and politically inappropriate to say anything negative regarding the fire service. The politicians' desire to associate with firefighters and not be critical can be advantageous, but nothing is automatic. Even though the importance of the emergency services and first responders (i.e., firefighters) has been reaffirmed, this has not always translated into more, better, or even appropriate resources to do the job. This is an area where you can determine what politicians really think. Watch their actions to see if they match their rhetoric. You might then deduce that more has to be done in the political arena to exert the necessary influence on the right people to obtain appropriate staffing, training, and equipment.

Of course, the fire and emergency services are not always on the front burner locally unless there has been a recent significant problem. Issues of importance to the voting public are related to good schools, safe communities, garbage collection, transportation, recreation opportunities, and other quality-of-life issues. Though there are exceptions (often related to emergency medical services, or EMS), rarely do fire and emergency services issues get people elected or re-elected, although they might cause someone to lose an election. Thus, rarely will one seeking election choose to do something considered anti–fire department. Can you think of cases in which the main issues in an election were related to the emergency services, whether fire, EMS, or another first-responder issue? Although you may know of some, certainly you can think of many more instances when crime, schools, public services, and other issues were debated and critical in an election.

So, if the fire service is not considered a major issue in the election arena, what can be done? There are two ways to approach the problem: either the fire and emergency service can become active in the political process or it can build relationships with those in power.

Local, county, and even state emergency services have sometimes exerted political pressure affecting an election. On a national level, the International Association of Fire Fighters has been an active supporter of political candidates. When their candidate is elected, things are good. When the opposition wins, there can be political paybacks. This approach requires knowledge of the culture of politics, whether local, state, or federal. Unless mandated by the appointing official (some chiefs are appointed by mayors), it is probably not a good idea for an individual chief to choose candidates to publicly support (unless of course they are certain to win!). Even if they are certain to win, it is important to think of the future. There will be other elections for the same (or even different) offices. Some jurisdictions have term limits. Bearing in mind that many politicians have long memories, the candidate who lost an election due in part to the chief's support of the opposition will most likely remember those who offered support when they do get elected!

If you as the fire chief cannot become involved in the direct campaigning, then the obvious choice of action is to develop good **political relationships**, that is, good dealings with those influential in the political arena. Remember, you do not need to compromise who or what you are. You can be true to yourself and still interact in the political arena while building relationships.

political relationships
Good dealings with those influential in the political arena

If you accept this as a vital role, you can be successful. You can probably identify many examples of such relationships. This is probably a safer route to take in the long run in contrast to being involved in the election process. The peril in direct political action is the consequence of backing the wrong candidate. Some communities change fire department leadership when elected officials change. The purpose here is not to debate the merits of such a system but to identify issues to be considered and preparations to be made.

What can or should you do? There are things to consider when developing your plan that may be effective regardless of the local form of government. These same tactics can be used on all levels of government—county, state, and federal—because they deal with relationships, not elections. The only difference is access and time as you get farther from local government.

GAINING INFLUENCE

First, you need to build a power base. How do you do that? Start with building relationships. Build relationships at *all* levels, *all* the time, with *all* the people. The political arena can be a very complex web where you never know who has access or influence. You may not always realize the relationships among people. "Be nice to all" is great advice and a rule of thumb to be followed by all members of your organization. You may discover a neighbor, businessperson, city employee, or other person seemingly out of the picture who wields clout in your community with elected officials (i.e., policy makers).

For example, you can build relationships in handling requests from politicians for your time and expertise. If you are asked for information, how do you handle the response? For instance, a local politician has been appointed to a committee within his professional association, and it happens to be in the area of first responders. He is preparing for the first meeting of the committee and wants to present well and establish credibility. You may be asked to help, and what you provide will help make his presentation a success. If you commit the time and energy, it will be well spent and will further establish you as the expert and create a bond for future dealings. How would you handle a request such as this?

Observe others who are successful, both in your community and your profession (other fire chiefs). What do they do that can be emulated? How did they get their power? How do they treat others? How do you treat people? Have you gained or earned respect from the community? You need to ask yourself these questions to understand where you stand. There is a reason why some people are successful and effective. They have learned the political process, invest the time, are very visible, and have the ability to get along with others. Learn the positive lessons from those who are good at what they do and avoid the errors made by those less successful.

You also need to know your business. You are the chief executive of the fire department, so you must have a solid knowledge base in your field. Although you must surround yourself with competent individuals because you cannot be an expert in everything, you will be expected to know what you are talking about. People will come to you for the right answers, so you must be prepared. Generally this is not a problem, but you must have accurate information. Keep current. Just as you tell firefighters never to stop training, so too you must continue learning. It is not possible to know everything in this ever-changing and complicated world. Besides continually educating yourself, know your resources. Where can you get answers and from whom? Your network of other experts, both within and outside your department, can be extremely helpful.

Dealing with politicians is often no different from working with others. Politicians want to be respected, treated fairly, and recognized for their position in the community. As fire chief you should do all you can to make sure they do not experience any surprises (a good idea in dealing with all your bosses). Do things to make them look good. Remember, they have an ego or would not have run for office. Control your ego so as not to clash with theirs. Do your job to the best of your ability and do not worry about their job. As mentioned earlier, their role is to set policy, and your role is to carry out that policy. Accept this. It is the way the system was established. Do what you can to influence policy, but in the end, remember what you were hired to do. You cannot have a hidden agenda or work behind the scenes against the wishes of your bosses.

You may occasionally get a chance to learn from those on the inside. If you can build a closer relationship with politicians, ask them for advice on how to work within the system. Regardless of your politics, you can learn from everyone. Remember that politics is not about Democrats and Republicans, donkeys and elephants, or red states and blue states but about various ways to be engaged in the political process. There are many resources available to help you. Keep your eyes open to all of them.

◆ INFORMAL POLITICS

The definition of politics as relations between people living in society refers to informal politics. This includes relationships with those inside the organization (e.g., employees), other governmental employees (local, county, state, and federal), the business community, and the citizens at large. There is a need to build relationships with all these groups long before issues arise. This relationship needs to be solid so the chief can consult or be consulted on major (and sometimes even minor) issues before it is too late. Politics will always be part of the future. Preparation for the inevitable is absolutely necessary.

Clearly, the fire chief needs to take a leadership and proactive role in building relationships with all who interact, or may interact, with the fire department and he or she must become involved in the formal and informal politics of an organization. However, this is not solely the responsibility of the chief. Others in the department need to accept politics as part of their role and must understand the department philosophy. The fire chief needs help in building relationships and must communicate to the employees what is needed and expected of them. Perhaps there should be a political component in every training and educational course offered. This would not necessarily need to be greatly detailed or complex but would include an explanation of the role politics plays in the organization. Department members would then be more aware of and better understand the impact of their actions (or inactions), though they might not always agree with the actions taken.

Obviously, politics is not only about dealing with politicians. Recall again the definition of politics as relations among people living in society. Now, think of all the people who have a role in providing emergency service:

- Labor (union)—employees
- Other city departments (police, emergency management if not part of FD, water, roads, building, engineering, payroll, finance, vehicle repair, maintenance, human resources, information technology, purchasing, legal, and others)
- Business community
- Citizen groups
- School districts and other educational institutions
- Bureaucrats on a county, state, and federal level
- Mutual aid and other fire departments
- Regulatory agencies
- Standard- and code-developing organizations (e.g., the International Code Council and the National Fire Protection Association)
- Professional associations (e.g., the International Association of Fire Chiefs)
- Hospitals
- Emergency medical service (if it is a private or outside service not part of the fire department or city)

You should be able to identify specific individuals or groups in each of these categories that are part of the emergency service system. You should make a list so that you are aware of all who contribute to the fire and emergency services in your area. You would be unable to do your job properly without input from these people. This is part of the complex society that makes up the world in which you operate (see Figure 2.3).

It should be obvious now that the fire department is not an island and that the fire and emergency service is truly a system. To be effective, you need to recognize everyone's role and build relationships so that the best possible service can be provided. There are those who contribute to your service delivery who are not within your chain of command. You must be able to exert influence when you do not have authority. You deal with other departments in your municipality, other governmental agencies, and other fire departments through mutual aid. You cannot relate to these people the same way as you do within your department. Work on this skill. It is very important. Failure to do so can negatively affect your department.

FIGURE 2.3 ◆ Others in local government are important to the success of the fire chief and the fire department.

Now, add to the normal considerations of politics the potential for jealousy. Imagine that you were politically successful in gaining a larger share of a budget increase than some of the other municipal departments. They might not think it was fair, and it could affect your working relationship. Bear in mind that the fire service relies on many to provide the emergency services. Working successfully with all these diverse people and groups is another example of the challenges that a fire chief and the department face.

◆ THE ROLE OF THE MEDIA

One element of politics that should be evident to all is the power of the media. It is a factor in so much of what is done in the world and plays a role in the formal and informal process. Good press establishes credibility, popularity, and can create power (e.g., the publicity following 9/11 enhanced the image of all firefighters). Negative press can lead to a downfall and can affect the image of the entire fire service or that of the local agency. It may have an impact on the image of the fire chief. This is very important, because the image of the head of an organization can affect the perception that others have on the entire organization—more than most people realize. If the chief is respected, then most likely the fire department is. When the chief has less than stellar opinions expressed regarding his or her performance, the entire organization can be painted with the same brush.

FIGURE 2.4 ◆ Your image and political interactions, both good and bad, can be influenced greatly by the media.

Media relations need to be considered as part of the entire political process. Of course, you as an individual can get too popular and challenge the egos of the politicians. Even if this occurs, you have options. If the fire department is popular, politicians have the opportunity to tout their role in its successes. The fire chief and the department will need to step back and let the political people get credit, whether or not they deserve it. There is a fine line unique to each community, dependent on the political system and the personalities involved. Take the time to learn where it runs in your community and the role of the media in affecting politics at the local level as well as at other levels. The media is a major contributor to community perceptions, so you need to work with it. People will see you as the face of the department, so make sure that this face is the one you want people to know (see Figure 2.4). Also remember that during the middle of a community crisis is not the time to start developing your contacts with the media.

◆ IN-YOUR-FACE POLITICS

Some people subscribe to the theory of confrontational politics, challenging those in a position of power. This can sometimes produce short-term wins but rarely works in the long run. You do a great deal of damage to relationships with this approach. People's memories are very long when they feel they have been wronged. You cannot embarrass someone into your way of thinking. It is not always easy, but you must seek cooperative solutions to issues. Whenever people are involved in issues there will always be an element of emotion and irrationality. Don't fight it; work with it. Do your homework, prepare, and be patient. It has been said that timing is everything. In politics this is very true.

◆ SUMMARY

Politics is a very important part of any job, but especially that of the chief executive of the fire department. If the main role of the fire chief is to obtain resources and use them efficiently and effectively, then relating to those who control the resources and determine their efficiency and effectiveness is extremely important. The following are some suggestions in this area:

- Recognize the importance of politics, possibly the most important part of the job.
- Identify what you need or should do, both formally and informally.
- Identify the key players—those with decision-making authority or those who influence those in power.
- Stay visible and participate.
- Control your ego.
- Study others who have succeeded.
- Use the media to build a base without becoming too visible.
- Know and maintain relationships with experts both inside and outside your department.

Politics is a part of everyday life. Everything you do contains an element of politics. Accept this fact and work with it so you can be the most effective fire chief possible. Building relationships that affect your politics is a long-term process that requires patience and endurance. Do not expect a quick fix, and do not get discouraged easily. Stay the course.

Activities

1. List events in your community in which participation would be beneficial.
2. List things you can do to build relationships.
3. Your local firefighters union backed a candidate for local office. The other candidate won. What actions would you recommend for the fire chief?
4. Identify and research a major political player in your community. How did he or she get power, how does he or she keep it, and how does he or she use it?
5. How do you deal with a split vote, a case in which one person controls the outcome, for example, a 4–3 decision? Identify strategies you can use to be prepared for a shift in the opposite direction.

◆ CASE STUDY

Politics and relationships will determine the success of the fire chief as much as any other part of the job. This case involves personal relationships and politics.

The mayor of your community has decided to run for state office, specifically for the House of Representatives. He has been a well-respected mayor and has supported the fire department. You have enjoyed a good working relationship and professional support. He has many supporters in the community including a majority of council members. Shortly after the announcement, the police chief for your community declares his candidacy for the same position. He is a personal friend of yours and has promoted a good working relationship between the police and fire departments. He promises to promote public safety issues if elected. Both people have called you and asked for your support and/or endorsement. Shortly after both candidates enter the race, your firefighters announce their support for the police chief. It is one year from the election.

What is your plan for handling this issue for the next 12 months?

CHAPTER 3

Human Resource Management

KEY TERMS

background check, p. 26	job description, p. 20	prerequisites, p. 20
hiring criteria, p. 19	physical ability tests, p. 25	probationary period, p. 27

If you ask fire chiefs what takes the most time on their job, they will most likely tell you "people" issues or problems. They will estimate that these constitute anywhere between 50 percent and 80 percent of their time. They may also offer that, either directly or indirectly, 20 percent of the people create 80 percent of the issues or problems (Pareto principle). Some may say that even fewer people create a larger percentage of the problems. Whether or not statistical evidence exists to support these concepts is irrelevant. What is important is that fire chiefs believe this to be true anecdotally and through their own experience; and, more than likely, the chief has not had significant preparation for handling these issues. Although procedures are changing, most people promoted have not had adequate experience or education to make them comfortable with this critical component of the job. It is all too common to take an excellent firefighter and make him or her a company officer. The knowledge and skills required to be an officer are different from those required to be a firefighter. (The same holds true for all positions—being good at your current job does not necessarily mean you have the skills for the next one.) By promoting someone without giving them the tools to succeed is a recipe for ultimate failure somewhere down the road.

Although it is important for fire chiefs to be prepared, it is also important for them to know where to get help. The world's knowledge base is expanding and changing so fast that all resources must be used. Legal issues, federal and state laws, local mandates, and city and fire department procedures all must be considered. Labor laws and labor relations are also a big part of human resource (HR) management. (This topic is addressed in Chapter 4.) As a result, human resource departments (sometimes known as personnel or human capital management departments) and legal advisors are valuable resources. They exist to help and should be considered experts in their field. They are more likely to maintain current information and generally provide good counsel. However, in most cases, the final decision resides with the chief. You gather the information, evaluate the options, and select the course of action. You cannot possibly know all the nuances, details, and specifics in all areas of

HR management. What you need to know are the basics and where to go for assistance. You will get the advice you need from the HR director or labor attorney, but ultimately, you need to consider all the input and make the final decisions.

The importance of HR management cannot be overstated. It takes significant time and is vital to maximizing the contributions of your most important resource—personnel. Although most of the following information appears to be about career departments, the same principles can be applied to on-call, part-time, and volunteer firefighters. The fire and emergency service is in the "service" business, which can be accomplished only with humans. Human resource management involves addressing interpersonal relations, group dynamics, team building and performance, and individual wants and needs while developing the important necessary teamwork. How individuals interact affects interpersonal relations and group dynamics. In the vast majority of cases, problems, concerns, and challenges are the result of personality conflicts. Recognizing this and resolving the issues takes skill, knowledge, and patience.

Providing for the individual involves being responsible from the time of his or her selection as a probationary employee to the day he or she leaves your organization. Ideally, individuals are hired, educated and trained, gain experience, get promoted, and ultimately retire. Human resource management is responsible for all the employee needs, wants, and desires from both the individual's perspective and that of management (with respect to the job). Of course, these services are provided within the context of what is good for the organization and the community served. The goal is to hire the most qualified individual, provide the resources to do the job (training and tools), create an environment that expects the best, and at the end of the career, allow a retirement that is long and prosperous. This is accomplished by taking the necessary steps to keep employees just as motivated throughout their career as they are on the very first day. This is quite a challenge for management.

◆ HIRING

Hiring good people is the first and most essential step. The saying that "you can't make a silk purse out of a sow's ear" is very true. Seldom do good people become bad, and vice versa. If you do not hire good people you will live with the mistake throughout their career and yours. Although people will gain knowledge and experience, they will not change their ethics, behaviors, attitudes, loyalties, and work habits without a tremendous amount of time and effort expended on the process, usually to no avail. These personal attributes are critical and must be considered.

You will first need to establish your **hiring criteria**, that is, the standards you will use to make your selection. Before developing your hiring criteria, though, you must investigate all applicable laws and ordinances. You need to know what you are mandated to do. There are federal and state laws prohibiting practices that can be interpreted as discriminating. Also, many localities have specific requirements. There may be civil service commissions or other established fire authorities that set criteria for hires. If so, they are the policy makers and you must play by their rules. You most likely will have input and therefore need to use your influence to make sure the process is designed to attract and retain great candidates. Whether there is a specific governing body that provides direction or simply an HR department, know the process and any nuances or special issues. There may be directives specific to selecting a workforce that is representative of your community. These can be established in

hiring criteria
Standards used in the selection of an employee

a formal process or even through a legal mandate (affirmative action) or by the direction of the policy setters in your community. Check them out ahead of time to avoid problems later. Examine what other organizations have done, including not only other fire departments but other agencies, particularly those you believe are good at attracting top candidates.

JOB DESCRIPTIONS

job description
A written document identifying the skills, knowledge, and abilities needed for a position

Although no hiring procedure is foolproof, there are things you can do to increase your chances of getting a great employee. First, create a **job description**, which is a written document identifying the skills, knowledge, and abilities needed for the position. For example, many departments are using firefighters to provide paramedic services, a service that wasn't always in their domain and that requires different and/or additional skills. If you use the same criteria for skills applicable only to firefighting, you are neglecting an important part of the job. Do not focus on one attribute, as you will need well-rounded personnel. For example, a paramedic license does not necessarily mean that you can discount other factors. Not all paramedics make good firefighters (and vice versa). Do not be tempted to save money in the short term for a possible long-term headache. Also remember that as the job evolves, so should your selection process.

Developing an accurate and applicable job description (also known as a *position description*) is essential not only for entry-level but for all positions. The job description puts into words your and your organization's expectations for performance and becomes the basis for your selection and evaluation. The description needs to be relatively dynamic, changing as the requirements and job duties change, so you should review it whenever you hire. Although there are many samples to copy, even one included here, do not be quick to adopt another job description without first evaluating your circumstances and fitting the job description to your department. The example presented in Figure 3.1 is in a format that can be adopted for all ranks. It is up to each department to fill in the information based on its specific needs. For other samples and additional viewpoints, work with your neighboring departments, your network, or HR experts. Seek out ranks that closely align with yours. For example, a captain in one organization may have job responsibilities similar to those of a battalion chief or lieutenant in another department. It is sometimes tempting to just substitute your information for someone else's in the belief that all departments are exactly the same. However, there are differences, so make sure to consider them.

PREREQUISITES

prerequisites
Conditions that must be met prior to application for a position

Be sure to consider any prerequisites for the job. **Prerequisites** are conditions that must be met prior to application for a position. Will you require any specific education (college), training (firefighter), licensure (paramedic), or experience? These can be important for many reasons, but especially because of the time and cost required to train the modern firefighter. Providing firefighter training and certification and EMS licensure can take more than 1000 hours, or approximately 6 months of work time. This is a relatively low estimate. If you require paramedic licensure in addition to firefighter I and II certification, the number of hours can easily exceed 1500 hours. Thus, you can hire persons and not see them perform outside of the training program for a large part of their first year on the job! Part of your selection process must consider all these factors. Will any of these circumstances have an immediate impact on

JOB DESCRIPTION
POSITION TITLE: ***Firefighter*** **SUPERVISED BY:** ***Battalion Chief***
DIVISION: ***Operations***
GENERAL PURPOSES
Works under the guidance and direction of an inspector or battalion chief.

ESSENTIAL DUTIES

1. Ensures that all assigned station apparatus and equipment are in a state of readiness as directed
2. Ensures that all subordinate personnel carry out their duties in a safe and efficient manner
3. Supervises subordinate personnel in their duties as directed
4. Coordinates, manages, and/or follows direction during emergency scene operations following established policy
5. Coordinates, manages, and/or follows direction during nonemergency station operations following established policy
6. Performs maintenance on department equipment, apparatus, facilities, and supplies
7. Participates in departmental activities
8. Ensures that his/her conduct and that of subordinates conforms to department standards
9. Reports violations of department policy and procedures to his/her supervisor
10. Attends and actively participates in training
11. Prepares reports that are accurate and complete
12. Maintains a positive working relationship with all department members
13. Coordinates and exchanges information with inspectors, other officers, and firefighters

PERIPHERAL DUTIES

1. Performs the duties of subordinate personnel and fills in for other staff members, as assigned
2. Attends meetings to keep informed of the activities of the department, the city, and the fire service

MINIMUM QUALIFICATIONS

Education & Experience

- Firefighter II
- Paramedic (AEMT) license with ACLS
- Certification at Hazardous Materials Operations level

NECESSARY KNOWLDGE, SKILLS, AND ABILITIES

Extensive Knowledge of:

- Department policy, procedures, personnel, facilities, apparatus, equipment and organizational philosophy

Thorough knowledge of:

- Fire behavior and characteristics
- Firefighting techniques, practices, and standards
- Fire inspection techniques
- Fire investigation techniques
- City and department policy and procedures
- Station apparatus and equipment capabilities
- EMS techniques, practices, and standards

Working Knowledge of:

- Department operation
- Department policies
- Skill in operation of listed equipment and apparatus

(*continued*)

Ability to:

- Work effectively with other staff, supervisors, and the public
- Effectively supervise subordinate personnel
- Follow verbal and written instructions
- Establish and maintain effective working relationships
- Handle the physical requirements of the job
- Analyze situations quickly and correctly and make decisions regarding the management of emergency situations
- Handle sensitive information in an appropriate manner
- Work effectively as part of a management team
- Identify problem areas and make recommendations
- Prepare concise department reports, correspondence, and records
- Communicate effectively, both orally and in writing
- Effectively analyze situations and provide solutions to problems

DESIRABLE KNOWLEDGE, SKILLS, AND ABILITIES

- Hazardous Materials Technician training

SPECIAL REQUIREMENTS

- Must maintain Paramedic (AEMT) license or higher
- Must maintain CPR certification
- Must maintain driver certification for all apparatus
- Must possess a valid state driver's license
- Must be able to read, write, and speak the English language

SELECTION GUIDELINES

May include any or all of the following: formal request for hire, review of education, training, and experience; written examination; oral board; background/driver's license check; hiring list; offer of employment; post-job offer physical examination, including drug screen

APPARATUS AND EQUIPMENT USED

All vehicles/apparatus, two-way radios, pager, personal computer, telephone, calculator, tape recorder, photo equipment, miscellaneous EMS equipment and supplies, operational tools, and related equipment

PHYSICAL DEMANDS

The physical demands described herein are representative of those that must be met by any member to successfully perform the essential functions of this job. Reasonable accommodations may be made to enable individuals with disabilities to perform the essential functions.

While performing the duties of this job, the member is frequently required to stand; sit; walk; talk; or hear; use hands or fingers, to handle, or operate objects, tools, or controls, and reach with hands and arms. The member is frequently required to climb or balance; stoop, kneel, crouch, crawl; and smell. The member must frequently lift and/or move heavy objects. Specific vision abilities required by this job include close, distance, color, and peripheral vision, depth perception, and the ability to focus.

WORK ENVIRONMENT

The work environment characteristics described herein are representative of those a member may encounter while performing the essential functions of this job. Reasonable accommodations may be made to enable individuals with disabilities to perform the essential functions.

Work is performed primarily in office, vehicle, and outdoor settings in all weather conditions including temperature extremes, during day or night shifts. Work is often performed in emergency and stressful situations.

The individual is exposed to sirens and hazards associated with fighting fires and rendering emergency medical assistance including infectious substances, smoke, noxious odors, fumes, chemicals, solvents, and oils.

The member occasionally works near moving mechanical parts and in high, precarious places and is occasionally exposed to wet and/or humid conditions, fumes or airborne particles, toxic or caustic chemicals, radiation, risk of electrical shock, and vibration. The noise level in the work environment is usually quiet in office settings, moderate during the daily work routine, and loud at the emergency scene.

The duties listed above are intended only as illustrations of the various types of work that may be performed. The omission of specific statements of duties does not exclude them from the position if the work is similar, related, or a logical assignment to the position.

The job description does not constitute an employment agreement between the employer and member and is subject to change by the employer as the needs of the employer and requirements of the job change.

Figure 3.1

service delivery? If so, you will have other issues to consider, further emphasizing the need for planning and anticipating needs.

An alternative hiring method that can save tuition costs and overtime, and allow for an earlier station assignment, is to require the appropriate education, training, and licenses. Very few occupations bring in raw talent without some level of preparation for the job. Could you imagine someone wanting to be a doctor without having the requisite education before practicing?

Establish your philosophy. Are prerequisites more important than other factors? Is it better to hire quality untrained talent and provide training versus saving time and money? You will live with the employee until you retire. Consider all the advantages and disadvantages and also consult the applicable laws to reduce the potential for creating an adverse impact on protected classes of applicants. Seek help and legal advice from the HR and legal departments.

THE PROCESS

Once you have identified the knowledge, skills, and abilities (KSAs) and the prerequisites for the job, you are ready to develop your selection process. Include everything you can think of that gives you the opportunity to identify the best candidates while operating within your established budget. You may have some restrictions based on what you can afford, but balance these against the need to make the correct choice. Hiring mistakes will cost more in the long term than your up-front investment if you don't get the right person. The following are some things to consider:

- Written test—to include analytical skills, critical thinking, mechanical aptitude, and writing ability
- Interviews—to identify verbal, analytical, and interpersonal skills
- Physical ability test—to identify whether the candidate has the necessary strength and agility
- Background check—to check references and previous employment and education
- Medical physical—may be given only after a bona fide offer of employment [per the Americans with Disabilities Act (ADA)] and should include a drug screen
- Psychological test—to identify the presence of the appropriate psychological background (also subject to ADA)

- Probation period—to evaluate performance. Organizations have used anywhere from 6 to 18 months as the length of probation. Generally, the longer the period, the more opportunity you have to observe performance and behavior. Consider it part of the selection process.

Each of these steps can be conducted by the department, community, or an unaffiliated third party. Many organizations use a combination. For example, there are companies that produce written tests. They are good, well written, and are generic for the position of firefighter. Few individual departments could develop a written test that would be adequate and also meet a potential legal challenge. You can rely on the nationally recognized test to begin the process of narrowing the field of applicants, but do not absolve yourself of all responsibility. Review the test prior to using it, considering the subjects and topics being tested, the reliability of predicting good candidates, and potential adverse impact. Don't automatically take the first test or the one that has always been used. The test should be up-to-date and relevant to your current circumstances.

INTERVIEWS

Interviews (oral interviews or oral exams) can be done by anyone who is qualified. A common belief is that anyone can conduct an interview, but this is not true. As with a written test, the first step is to identify the areas to be evaluated and then to develop questions with their correct answers in mind. Individuals used on an interview panel must be trained in the process, the legal issues, and the scoring system to identify desired candidates. Properly conducted interviews are valid and successful in predicting qualified individuals but only if the panel members are adequately prepared.

Formal training is suggested, as well as a preinterview meeting to discuss the situation at hand. This does not mean that candidates should be preselected or that you should hint as to who might be preferred. Rather, the meeting is to make sure that all interviewers understand the process and their role. The panel may include individuals from inside or outside the department, or both. Another option is to conduct two separate interviews, one external and one internal, to obtain different perspectives and, it is hoped, to corroborate findings. Panelists may come from within the department, other fire departments, other city agencies, the HR department, or other places that have an interest and understanding of the requirements. It may also help to select panelists from among supervisors (those who will be responsible for the performance of the person hired) and perhaps a labor representative. It is most important to make sure all the interviewers are qualified and prepared.

One thing to remember when conducting interviews is to ask open-ended questions. Avoid asking questions in such a way that they elicit a yes or no response. Further, ask for examples. For example, if you ask interviewees whether they are hard workers, they will say yes. The proper way is to ask for examples of things they have done that demonstrate they are hard workers. Be prepared with follow-up or probing questions based on the answers given. Many candidates are prepared for the first line of questions. Your second set should force them to think on their feet.

You can get good interview questions from a variety of sources. Try contacting your network of experts, or search the Internet for resources. HR directors are in the business of interviewing and should be able to give you helpful information, and some consulting firms may be willing to provide assistance. You can also craft questions based on your own experiences. More than likely you have been on the receiving end of interview questions. Try to recall the ones that you thought made sense. Volunteering to help other departments with their process will help you gain experience both by doing and by learning from others who sit on panels more frequently than you. Finally, look for seminars and workshops on interviewing offered by private entities, associations, or governmental bodies. As with any other endeavor, training and practice will make you better and more comfortable with this aspect of the hiring process.

PHYSICAL ABILITY TESTS

Physical ability tests are administered to determine the candidate's strength, endurance, and agility with respect to job performance (see Figure 3.2). This step in the process is the one most often challenged with respect to adverse impact. Many different tests have been used over the years. Often, they have been developed locally or adopted from other departments and have been based on either individual or group experiences and beliefs as to what it takes to be a firefighter. In some situations consortiums have attempted to agree on a valid test, one that is legal and is good at predicting positive outcomes. In 1996 the International Association of Fire Chiefs (IAFC) and the International Association of Fire Fighters (IAFF) joined forces to develop a test that could be used to reasonably predict successful candidates. Ten metro departments met and ultimately released the Candidate Physical Ability Test (CPAT). The two organizations also asked the United States Justice Department (which typically has jurisdiction in cases of alleged adverse impact) for its opinion. The Justice Department gave generic approval. However, this does not relieve fire departments of their responsibility to evaluate the applicability of CPAT to their jurisdiction. Regardless of the test used and its origins, it is up to the local department to validate the test as to its application in the selection process.

physical ability tests
A series of tests administered to determine a candidate's strength, endurance, and agility with respect to job performance

FIGURE 3.2 ◆ Agility testing is an important part of the selection process.

BACKGROUND CHECKS

background check
A look into the history of the candidate relevant to the position being sought

A **background check** is simply a look into the history of the candidate relevant to the position being sought and is a must as you look for the best person for the job. Past performance is often the best indicator of what may be in store in the future. Lessons may be learned from law enforcement agencies, which commit significant resources to background checks, because they reveal a lot of information regarding the applicant. They are also expensive, which is why many fire departments shy away from using them—but what is the cost of a hiring mistake? Whether you use a background check or not should be a conscious decision based on its usefulness compared with the cost. If budget considerations are a significant reason for limited background checks, use the parts that are relatively cheap and easy. Always do a criminal, driver, and sex offender check and have established criteria as to what is and isn't acceptable. For example, is there a difference between a felony and misdemeanor? How many points on a driver's record are okay? Do certain violations mean automatic disqualification? Will you check extenuating circumstances? You need to determine these criteria in advance to avoid the perception that candidates are treated differently or unfairly.

Besides performing these basic checks, call past employers and/or others with knowledge of the candidate, but first, obtain a signed waiver from the applicant. Although this does not eliminate all risk, it greatly reduces your liability. The most important question to ask is whether the individual would be rehired given the opportunity. Also, you may need to read into any recommendations you get. Some people have been challenged on their evaluations and will not go to extremes, either positive or negative. They may be hesitant owing to prior challenges to their recommendations. Use your instincts and read not only what is said but how it is said. Remember, the background check is just part of the hiring process and must be weighed with other components.

PSYCHOLOGICAL TESTS

Psychological tests have become more popular in recent years as departments attempt to identify characteristics present in candidates that are relevant to firefighters. Such tests have been utilized more extensively by law enforcement, but they may be another tool in the selection process. There are psychologists and other mental health professionals who have the ability and training to conduct such tests. The purpose is to identify both good qualities and issues that may send up a red flag. Do your research to find qualified individuals to perform these tests, remembering that they are just another factor for you to consider. Such testing may be considered a medical procedure and must comply with the Americans with Disabilities Act, meaning that it can be administered only after a job offer has been made. Consult your legal advisor and HR department for clarification and advice.

PHYSICAL EXAMINATIONS

A physical exam by a licensed, competent, and occupationally trained physician is a must. The examination must be conducted based on the requirements of a firefighter. The National Fire Protection Association has an established and recognized standard, NFPA 1582, "Standard on Medical Requirements for Fire Fighters." The physician conducting the physical must be familiar with the standard and your job requirements. Generally, the doctor will not provide specific results but only a recommendation to proceed with the hiring or a caution that the candidate is not physically able. As with the psychological test, review the requirements of the ADA. The medical exam may be administered only after the job offer has been made. You may also wish to include a drug screen for all applicants. Check appropriate laws, but remember there may be a difference between those for new hires and incumbents.

PROBATIONARY PERIOD

The last, and maybe most important, part of the selection process is the **probationary period**, which is defined as a defined period of time during which a person's behavior and ability is observed and tested. The length of time for probation should be 12 to 18 months (the longer the better). This is your chance to see if the candidate has what it takes to be the firefighter you expect. Usually, probationary employees are "at-will" employees, meaning they can be dismissed without cause; however, this does not mean you should not have a reason. Give everyone a fair chance to succeed. If it is evident there are issues, do not be afraid to dismiss the employee. Seldom does the effort get better, and problems during probation rarely go away. Many a chief has had to live with an individual who slid through probation when indications were obvious that there were issues with his or her performance. You need to make the tough choices early in someone's career.

probationary period
A defined period of time during which a person's behavior and ability is observed and tested

Of course, you as the chief rarely have the day-to-day contact with individuals necessary to make a proper evaluation. The supervisors need to accept this responsibility and meet the challenge. Your job is to prepare them to perform the evaluation and to make sure they understand the importance of their role in the selection process. It is in the interest of every department member to hire competent, qualified individuals. Make sure your supervisors know what is expected and understand their responsibility to properly evaluate all candidates. Also understand that your supervisors will have their own biases based on their interaction with the employees. They may also have some hidden biases. Make sure your supervisors know the importance of treating others fairly and appropriately. Remember, it is you, not the supervisor, who will be called on the carpet to explain disparate treatment. You will have to defend your actions and those of your subordinates. Your job is to give the supervisors the tools and make sure the candidates get a fair shake.

All recruits must have a clear understanding of your and the department's expectations. These must be communicated directly, and the probationary employee must acknowledge what they have been told. New employees are very excited, so they may not always understand or remember. You should consider a form that explains the requirements and allows for the probationary employee to verify receipt of this information by initialing as the explanation is made. Figure 3.3 is a suggested form. You would need to add the elements important to your organization.

Establish regular formal evaluations during the probationary period. Develop forms that clearly identify the expectations during this phase of the evaluation period, which is part of the selection process. What does the candidate need to know at 3, 6, 9, and 12 months? Make this as objective as possible, but realize there is a subjective component to the evaluation period. Rarely should a candidate make it to the end of the probation and be dismissed. There should be a record of unsatisfactory performance and clear instruction about what is to be improved. The consequences, that is, dismissal, of failure to meet your standards must be clearly stated.

As you develop your hiring process, remember to include the experts, those in the personnel or human resources department. There are many nuances to hiring. Others may do it more frequently than you, so they have more experience and insight. This does not suggest that you should be absolved of all responsibility. Many a fire chief

SAMPLE PROBATIONARY REQUIREMENTS ACKNOWLEDGMENT AND EXPLANATION

Having successfully completed the hiring and selection process, you are now appointed to the position of probationary firefighter. Below are listed various items relevant to the probationary requirements and the expectations of this department. These items will be reviewed with you and you will verify that you have received an explanation and you understand each of the items. When the items are being explained, pay attention, as you will be expected to initial next to each statement that you understand what you are being told. You will return the completed form, and it will be placed in your personnel file.

INITIALS	REQUIREMENTS
________	I have been given the recruit handbook and am responsible for reading and understanding its contents.
________	I will conduct myself in accordance with the rules and regulations at all times.
________	I am responsible to follow all oral and written direction given by the department, its officers, and representatives.
________	I will complete recruit school with a minimum grade of 80%.
________	I will become licensed as a paramedic prior to completion of my probationary period.
________	The probationary period is 18 months.
________	I will be evaluated every 3 months. Unsatisfactory reviews will be grounds for dismissal.

Note: This is a sample. You are encouraged to add to it to create your own list based on the requirements of your department.

FIGURE 3.3

has been saddled with a less than desirable candidate because he or she was not involved in the process. Thus, it is extremely important to use the many resources available to you. You want the best candidate. People rarely change after they are hired. Often, you will hear that current "problem" employees were always that way. They had issues from day one. Unfortunately, in too many cases, the completion of probation means that you are stuck with a less than stellar performer for their career, so take full advantage of the probation period to thoroughly assess candidates.

TRAINING AND ORIENTATION

Once individuals are hired, evaluate then with respect to their education, training, and experience. What do they know and what will they need to know to begin the job? The orientation introduces new employees to the information needed to begin work. They can't get everything in the first day, but they need enough information to get started. As for training, a suggestion is to develop a checklist for the orientation period as well as the entire probation period, because the individuals may have had training that does not meet your expectations with respect to procedures, equipment, and personnel. The checklist will serve as reminder of all the topics that need to be covered, ensure consistency among recruits, and provide a permanent record for the individual. It will also aid your supervisors and promote consistency from shift to shift.

First, identify the tasks that the new employee needs to know, then specify the timeframe in which they should be presented and mastered. For example, firefighters need to be able to operate with a self-contained breathing apparatus and specifically the make and model used by your department. That would be a task in which the new employee would need to be proficient before being assigned to the apparatus. This task would typically be presented and mastered near the beginning of the training program, most likely on the first day. Even though they may be from recruit school, the officers must be confident that they have a sound foundation. Another task might be to have a good working knowledge of the street system. It might not be necessary to know this on the first day, but there should be a deadline for mastering this task. Your job (or that of whoever gets the assignment) is to identify what is to be known, who is going to present it, how it will be evaluated, and how long it will take.

Obviously, those without any training need to attend basic training to include fire, EMS, and other core functions. In addition to having the same basic training, recruits need to continue to learn more about the job during their orientation. A program for new hires must include information about such items as policies and procedures (department and city); federal, state, and other mandated requirements; EMS items (which may include health issues such as vaccinations); assignments; and specific job expectations. Again, identify tasks that they are expected and need to know and set a timeline for their mastery.

Recruits should be assigned to existing firefighters who can serve as mentors, coaches, and one-on-one instructors when they are finished with basic training and assigned to a station. Choose these mentors wisely. Select those who are interested, have the capability, and set a good example. Much of what a recruit will retain will be what they learn from their exposure to this experienced person. Make sure that mentors understand their role, receive training for their role, and are empowered to carry out the assignment.

Annual evaluations are important and will be covered later in this chapter, but during the first year there should be at least three reviews. A suggested schedule is at 3, 6, and 12 months. If you have an 18-month probationary period, extend the intervals. This gets a formal evaluation on the books and lets the probationary employee and the department know where everyone stands. If adequate progress is being made, all is well. If not, there is time to make corrections at each review or to decide that the employee is not going to make it. There should be an evaluation form with very specific requirements outlined. This removes all questions as to expectations and performance. Remember throughout the probationary period that it is part of the selection process. Get firefighters to develop good work habits. Those that can't, or won't, meet your requirements need to be dismissed.

◆ CONTINUING EDUCATION AND TRAINING

How do you maintain skills throughout a career and also prepare individuals for future promotions? Obviously, through training, education, and experience. With the exception of small organizations, the fire chief cannot do this directly. The job of fire chief is to establish a system that will give members the tools to perform their job. Some things are the responsibility of the employee, and others are better handled by the department. Training programs are needed to provide required continuing education (mostly needed for EMS licensure), to maintain a high level of competence in core responsibilities and skills, and to introduce new ideas and programs to the employees (see Figure 3.4). In today's rapidly changing world and with additional job

FIGURE 3.4 ◆ Classroom instruction is important in the development of personnel.

requirements, this can be a challenging assignment. You need to assign it to others in the organization, but you need to maintain oversight and review. There are many resources to help you. Check with other departments and professional associations.

◆ PROMOTIONS

Future officers almost always are promoted from lesser ranks. As a result, you will need to promote based on ability and skills. Do not be bullied into selecting someone who is not prepared or qualified. The department needs to help in preparing future officers. Few occupations promote individuals and then let them learn on the job. Members need to know the requirements of the next position and take the necessary steps to prepare for it. The department can help by providing mentoring and the necessary applicable training, including courses certified within your state or those that have relevance to your organization. This training needs to continue until the individual retires.

The organizational structure of the fire department requires officers in various positions, so promotional processes are needed. They must be developed to be fair to those seeking promotion and also must help in selecting the best candidate. This process is somewhat similar to that of hiring a new employee: identify the skills, knowledge, and abilities needed, and select tools to determine the best candidates. The tools can include written and oral testing. Department evaluations and recommendations for senior members should also be considered. Remember, the best firefighter does not always make the best company officer; the best company officer does not make the best command officer, and so on. NFPA 1021, "Standard for Fire Officer Professional Qualifications" can be used as a guide to help determine qualifications and prerequisites for promotion to company officer. As for any other position, a job description is necessary to help to focus on the qualities needed.

You must know whether there are any mandated promotional requirements, for example, through a labor agreement (collective bargaining agreement), local ordinance, or civil service commission. Check your local requirements. There may not be any choices under current agreements. If this is the case, you have little to do, but you should investigate and determine whether the current system is working. Many have been in place for a long time and have not been reviewed. If you find the system needs amending, find out how to do that, because poor promotions affect the organization for a long time.

After you have tallied all the scores and considered all other components of the promotions, you need to make the choices. Generally, there will be more people upset that they did not get promoted than happy ones because they did. One way to address this situation head on is to deliver the message directly not only to those who got promoted but those who didn't. Although few people believe they were not the best person for the job, most appreciate direct communication from the boss regarding their status. They may have a few questions and will be grateful for any time you can give them.

After an individual gets promoted, what happens? Do you have a system of orientation, training, and education? If not, you need to establish one. Remember, people most often perform to their expectations, so you need to let the new officers know what they are. They are the ones who carry your message. How well they do that will determine the buy-in from the rank and file.

New officers need to be prepared to do their job, and they need to be evaluated. A probationary period for new promotions is a good idea, as it is part of the promotion process. It gives you the chance to see the individual perform. Coach and correct for desirable behavior. Good officers are not born; they need assistance and mentoring, either from you or another competent officer, in fulfilling their potential.

Remind the officers that they asked for the promotion and were not forced by you to accept it, so they need to accept the responsibilities that go with the rank. Be direct with the candidates so they understand your position and your expectations of high performance. Remind them that they must enforce rules, policies, and procedures regardless of their own beliefs on the issues. Transitioning from firefighter to supervisor (or any other promotion), from buddy to boss, is very challenging. With every promotion, more performance is expected and more support of and loyalty to the administration. Communicate all this to your officers.

◆ PERFORMANCE APPRAISALS

Doing performance appraisals on everyone in your organization is not a fun part of the job. Of course, depending on the size of your organization, you don't need to do them all. They should be done by the direct supervisor. The challenges of appraisals are twofold: no one likes to get the news that they are anything short of an

FIGURE 3.5 ◆ Proficiency in the basics must be tested on a regular basis.

A student, and those doing them are usually ill-prepared and do not like that part of the job. Nevertheless, appraisals have become more important and can be good for the organization if handled correctly (see Figure 3.5). Review some of the literature on the topic and consult with your HR specialist to get a handle on it. In general, the appraisal is based on the job description, so have a current one on all positions. Develop a form specific to the position being appraised. It should include the job essentials and can also include opportunities for the supervisor to offer an opinion on performance. This should be substantiated from the perspective of the supervisor. Lastly, provide basic training to your personnel in performing the appraisal.

◆ SUMMARY

Clearly, personnel are the key to a successful fire department. The concept is relatively simple: hire good people, provide them good training, keep them motivated, promote the best people for the job, and take care of their needs. Remember, personnel issues take a lot of time. The better you are at addressing them, the less time it will take. Commit the necessary time up front to avoid significant issues later. Communication, is a key to being successful, as is seeking the advice of your local experts who deal with these issues more regularly than you do.

ACTIVITIES

1. Locate the Family Medical Leave Act policy for your community. Assume one of your members has requested leave under this act. List the actions that you would take.
2. One of firefighters reported 30 minutes late for work. Discuss how you would address this situation.
3. A female firefighter approaches you to tell you she believes she is being harassed and that there is a hostile work environment in her station. What would you do?
4. Determine the hiring procedure for firefighters in your community. Then, discuss whether it needs to be changed to better reflect the current job functions.
5. While on vacation, you receive a call that one of your firefighters has been arrested. What would you do?

CASE STUDY

The culture of the fire service is changing—reflecting the current society. The differences between generations can challenge the fire chief with respect to human resource issues. As fire chief you know that your organization requires uniformity (as much as possible) and a mainstream appearance to the public. Personal appearance has become an issue in your organization. More firefighters are sporting more visible and colorful tattoos. The members are also showing more jewelry and body piercing. Some members of the department and community have noticed and wonder what your policy is. Currently, your department has none regarding tattoos.

What will you do?

CHAPTER 4

Labor Relations

KEY TERMS

binding arbitration, p. 43
due process, p. 37
fact finding, p. 42
grievance, p. 45
International Association of Fire Fighters (IAFF), p. 35
labor agreement (contract), p. 36
mediation, p. 42

Labor relations are a critical component of the fire chief's job. In this chapter we discuss human resource management but in the specialized area of dealing with employee groups that are organized as unions. Of course, good labor–management relations work in nonunion environments as well. Establishing relations with labor organizations may be regulated in some states, so you are advised to research any applicable laws or rules in your region. Regardless, working with a union to build strong relationships is vital to the success of your organization. These relationships take time to build and work to maintain, but they are worth the effort, especially because they help maintain morale and prevent wasting time on issues that do not directly relate to your function to serve the public. You would be hard pressed to find a well-run fire department that has contentious or confrontational relations between labor and management.

Because people are your most vital resource, you should desire motivated employees. Recall the first day of new employees. They are excited and willing to do just about anything. Although they may not have the skills and abilities to perform at the expected level, they are highly motivated to do the best job as firefighters. Subsequently, if they are treated well, they will remain motivated, and this motivation will translate into better service to the community. However, if they believe they are not getting the treatment they deserve, they will not perform at peak proficiency, and morale will suffer.

Time spent on labor issues that could easily be resolved is time away from the core mission of the department. Cooperation is the most appropriate course of action within the organization. Although, this is not always possible (mostly because of conflicting egos and personalities), a true leader, that is, the chief executive of the fire department, will understand the value of good relations and will do all in his or her power to promote an excellent work environment. This may not seem possible owing to outside forces, but the effort to work within the established guidelines must be put forth.

FIGURE 4.1 ◆ The International Association of Fire Fighters represent the vast majority of career firefighters.

◆ THE BASICS

Labor unions were first organized because workers believed they were not being treated fairly. In the beginning disputes were mostly over wages, benefits, and unbearable working conditions. As these issues were somewhat resolved (they may never be fully resolved, as everyone believes they are overworked and underpaid) the roles and responsibilities of the union expanded to other areas. Whatever the issue, the union movement is about organizing to fight against unfair treatment (or perceived unfair treatment).

It should be noted, though, that the fire service is unique. The major representative of firefighters in the United States and Canada is the **International Association of Fire Fighters (IAFF)** (see Figure 4.1). Although it is organized to help locals improve wages, benefits, safety, and working conditions, the IAFF is as much about the brotherhood (and now sisterhood) as it is about traditional labor issues. There is a strong draw to the IAFF based on this need to be part of the larger group. Even if the employees are treated well and fairly, they still need to be part of the labor movement.

International Association of Fire Fighters (IAFF) The major representative of firefighters in the United States and Canada

◆ RESPONSIBILITIES

One of the important steps toward working with labor unions is to understand their function and responsibilities. Their role is to represent the workforce as a whole while defending the rights of the individual within the context of the labor agreement (contract) and the law. Typically, unions negotiate items related to wages, benefits, working conditions, and safety. In some cases working conditions and safety have been liberally interpreted to allow discussions on a wide variety of seemingly unrelated issues. In other situations negotiations have tended to remain more conservative along the lines of traditional issues. The reasons for this difference relate as much to the culture of the community and the leadership of the department and union as they do to established rules. Communities with strong unions tend to have a wider range of topics, and communities with strong management have a more restrictive view of what should or shouldn't be included. State laws also have an impact. Right-to-work states may have laws that allow less interaction between labor and management and may even restrict membership. Some states may prohibit official contact between management and a sanctioned union. Again, know what is allowed in your state.

If good relations are so important, why is that not the case in many departments? There can be a wide range of reasons. It could be that the culture of community does not encourage cooperation, or the union may believe that it needs conflict to be perceived as effective. Sometimes the members wonder why they are paying dues if they perceive that the union is not fighting for every last benefit available. In other cases there have always been contentious relations and the players don't even know why; it has just always been that way). Sometimes the chief and the union president just don't get along, perhaps because of clashing egos or personalities. The problem might even reside outside the department and within the administration of the municipality, outside the control of the fire chief. If labor and management are cooperating, the chief's boss or members of the labor group might think the president and chief are "in bed" together; that is, there must be something going on with the relationship that is not conducive to management or labor. The suspicions can be difficult to address. Not everyone embraces cooperation. In most cases, the root cause of poor or ineffective relations is a breakdown in communications, poor interpersonal relations, and a failure to understand the respective roles.

In some communities, labor unions have become very powerful because of successes in politics. Local unions that have been aggressive in establishing good relations with local politicians usually reap benefits. They establish the relationship through campaigning, donating to campaigns, delivering votes, or through personal relationships. Regardless of how the union obtained its power, the fire chief needs to know its level in the community. Because of successes in the political arena, many locals have a great deal of influence on the direction of the fire department, sometimes more than the fire chief! There are many examples of fire chiefs who did not acknowledge this fact and had either a short tenure or found it very difficult to accomplish their goals. Many an intelligent chief, someone who knew and understood the ins and outs of running a fire department, was not able to lead the organization effectively because he or she did not understand the power base of the labor union.

◆ PRIMARY ROLES

labor agreement (contract)
A formal agreement on wages, benefits, and working conditions (mostly those related to safety)

You should learn as much as you can about labor relations, especially on those issues most important to your local situation. A basic understanding of the role of the union is a must. Although organized labor can have other ancillary functions dependent on circumstances within the community, its primary role is to

- Negotiate a **labor agreement (contract)**, the formal agreement on wages, benefits, and working conditions (mostly those related to safety);

- Meet and confer on issues affecting the union;
- Address safety concerns not in the labor agreement;
- Ensure that the contract is enforced appropriately and consistently;
- Ensure that **due process** is followed when discipline is administered to members of the bargaining unit, that is, that formal proceedings are carried out in accordance with the established rules;
- Represent the entire union or individuals in cases requiring mediation and arbitration.

due process
Formal proceedings carried out in accordance with the established rules

Of course, there can be conflict within any organization when labor and management do not agree on their roles. Clearly, a good understanding of the function of the union is the first step toward establishing good working relations. For example, you should know that the union generally is responsible not for defending the actions of individuals but for protecting their rights as outlined in the contract or law. Often, conflict arises when an individual is not afforded their due rights. Another important concern is equal treatment. One of the basic functions of a union is to ensure everyone is treated fairly, that is, equally. If an individual is disciplined for an infraction, the punishment must be consistent. Further, this means not deviating from the labor agreement even if it means assisting a firefighter. For example, if the labor agreement specifies leave requirements, do not interpret them more liberally for certain individuals, as this could set a precedent and might also place an individual in a position of being a favorite, with special privileges.

As fire chief you may sometimes be caught in the middle between labor and your bosses, that is, the management of the city or community. Your background in the fire service may make you sympathetic to the issues being presented; however, management may expect more support for the community position. You often walk a fine line between being too close to the union or forgetting your roots, yet sometimes you will believe in management, and other times you will feel obligated to support the union position. For example, you may be aware of contract language that others outside the fire department may not know. If they have authority over you, they may expect 100 percent loyalty. You need to know your limits and when to choose the appropriate side according to the circumstances. This is another instance in which relationships are very important. Building them before an issue arises is good practice.

THE CONTRACT OR LABOR AGREEMENT

Negotiating for a contract between the governing body and employees is not the most enjoyable part of the fire chief's job. In spite of everyone's best intentions the process is often not as cordial and impersonal as some would like to believe. It is important for you to stay out of the firing line as much as possible. If you do not control the finances or other budgetary items, then there is no reason to become involved in that part of negotiations. Your role in negotiations should be on issues that affect the operation of the department, such as scheduling/time-off issues, promotions, and other mostly nonfinancial issues. Of course, you want to make sure your employees are fairly and competitively compensated, as it makes it easier to recruit and retain employees. If the community does not compensate its employees relative to the local market, then quality candidates will search elsewhere for employment.

As much as you would like to stay totally out of the negotiating process, the reality is that your involvement is needed, as decisions might be made that could make it difficult to manage the department. For example, a negotiator for the community may not understand the 24/7 operation of the fire department and may offer more time off to reduce costs. You may need to point out that minimum staffing requirements might necessitate callback of personnel for overtime, which would affect the budget. Once the contract is signed, you must work with the agreement, but when the overtime budget is exceeded, some people may not remember the negotiations. You may be instructed to reduce the costs, even though you may not be able to do so. The time to raise these issues is during negotiations. For this and other reasons, you need to be involved.

Being prepared for negotiations is important. Find out and agree to your role before they get started. As much as you would like to stay neutral and keep things on a professional level, you will be challenged to do so, as negotiations can be as much emotional as they are logical. You should have answers to the following questions before you begin negotiations:

- What issues are important to the fire department? Remember, you are in the middle on some issues. You may need to fight for things that help you maintain a good working relationship with the firefighters and not do anything that will damage morale in the long term.
- What issues are important to the city? These can be the big-ticket items that affect all labor unions in the city, such as pay increases, pension benefits, and health care. All these are important to all employee groups. The community will be conscious of the potential impact of these items in future negotiations with other labor groups.
- What issues are important to you with respect to administering and leading the department? A good management rights section of the contract is important to protecting your rights.
- When should you side with the boss? Although you are encouraged to stay neutral whenever possible, sometimes you need to be a team player and show your loyalty to your boss. As much as possible, know what these issues are and how they may be addressed.
- When should you stay neutral? It is generally best to stay out of financial areas, for example, pay and benefits. Usually, this is easy to do, because often pension and health care applicable to firefighters are also part of the fire chief's package. Those negotiating on behalf of the community usually understand this and do not place the fire chief in a precarious position. After all, you also want a pay raise, better pension, and excellent health care. The union may be negotiating your next pension enhancement!
- When should you push for the local union? Many labor leaders can be reasonable and prudent on issues affecting the community and the organization. They may understand the financial constraints based on the economy (though some management personnel would say this never happens). They may also have some suggestions that could positively affect the way you do business. Do not discount this. Keep an open mind and when appropriate support the union position.

Remember as you negotiate that those who must live with the provisions of the document will have much more time with it than you do. Often, the chief has limited time left in the department. The contract is a legacy for those who follow. You are dealing not only with your own issues but maybe those of your successor. Also bear in mind that there are many chiefs who negotiated a deal while union officials and now have to administer it as chiefs! (See Figure 4.2.)

SAMPLE MANAGEMENT-RIGHTS CLAUSE

The City Council on its own behalf and on behalf of its Electors, hereby retains and reserves unto itself, all powers, rights, authority, duties, and responsibilities conferred upon and vested in it by the laws and the Constitution of the State of Michigan and of the United States. Further, all rights which ordinarily vest in and are exercised by employers are reserved to and remain vested in the City Council, including but without limiting the generality of the foregoing, has the right:

(a) to manage its affairs efficiently and economically, including the determination of quantity and quality of services to be rendered to the public, the control of equipment to be used, and the discontinuance of any services or methods of operation;
(b) to introduce new equipment, methods, or processes, change or eliminate existing equipment and institute technological changes, decide on supplies and equipment to be purchased;
(c) to direct the work force, to assign the type and location of work assignments, and determine the number of employees assigned to operations;
(d) to determine the number, location, and type of facilities and installations;
(e) to determine the size of the workforce and increase or decrease its size;
(f) to hire new employees, to assign and lay off employees;
(g) to establish and change work schedules, work standards, and the methods, processes, and procedures by which such work is to be performed;
(h) to discipline, suspend, and discharge employees for cause;
(i) to determine lunch, rest periods, and cleanup times, and the starting and quitting times;
(j) to subcontract or purchase the construction of new facilities or the improvement of existing facilities;
(k) to subcontract or purchase any work processes or services in line with past practices;
(l) to select employees for promotion or transfer to supervisory or other positions within the department;
(m) to establish training requirements for purposes of maintaining or improving professional skills of employees. The City reserves the foregoing rights except such as are specifically relinquished or modified by the terms of this Agreement. It is agreed that these enumerations of management prerogatives shall not be deemed to exclude other prerogatives not enumerated, and except as specifically abridged, delegated, modified, or granted by this Agreement, all of the rights, powers, and authority the City had prior to the signing of this Agreement are retained by the City.

FIGURE 4.2

You should have some general background on negotiations. There are many resources you can tap, such as books, seminars, and workshops. Finally, recognize the experts who work for your community. The labor attorney and human resource professional are invaluable because they generally have much more experience than you do with respect to contract negotiations. They negotiate with many more labor unions and maybe more communities (if the labor attorney is a contract employee), so they may know more trends and keep more current. Consider them your most valuable resource when learning about this part of the job.

Know your role with respect to your labor attorney and HR director. During the best of negotiations, relationships will be tested. The labor attorney and HR director do not need to return to the fire station, work with firefighters every day, and make sure the provisions of the contract are carried out. They don't need to worry about long-term relationships because they don't work with the firefighters every day. You need to attempt to keep some distance from the issues that can cause hurt feelings while maintaining loyalty to and support of the management position. It is not an easy task. You need to acknowledge it and develop strategies to address these concerns.

PREPARING FOR BARGAINING

Although the subject of negotiations and bargaining is too complex to adequately handle here, the following are a few things you should remember as you approach any bargaining situation:

- Know what you want and don't want.
- Get everything out on the table as soon as possible. You do not want to introduce new ideas and concepts late in the process.
- Be prepared. You do not want to be surprised, and you need to know more that the others with whom you are negotiating.
- Be patient. Seldom are things settled in one session. Negotiating involves two parties who both need to be thinking. Plan for the long haul.
- Remember relationships. If you know the players, you can better understand their needs and desires. Good interpersonal relationships lead to better agreements.
- Be aware of emotions. They can affect logical thought processes. Emotions can affect perceptions and be contrary to the facts.
- Do not hesitate to call for a caucus with your negotiating team if you have an issue that requires a private discussion. Avoid saying things in a public forum that should first be discussed in confidence with the negotiating team.
- Maintaining your sense of humor can help.

ADMINISTERING THE CONTRACT

Once the contract is settled, it is up to you to manage and administer it (see Figure 4.3). Too often a fire chief has neglected to consider the content of the agreement when making a decision, and the decision (even though it might have been correct) was overturned. You must adhere to the provisions of the labor agreement whether you like it or not. The following are things that you should do:

- Read the contract. Know what it contains and what general topics are covered.
- Keep the contract handy so you can refer to it.
- Get interpretations when needed. Not all sections will be crystal clear, and some parts may be subject to interpretation. Although you don't always need to get a legal opinion, there may be occasions that warrant consultation with your department's legal council. You will also want to know what your boss thinks. It is easier to adjust your views early in the process based on this perspective as opposed to having a grievance filed and your decision being overturned by your boss. You want to take the position that management takes.
- Follow the rules, like it or not.
- Make notes as to sections that need revision in the next negotiation cycle.

SAMPLE ITEMS IN A LABOR AGREEMENT

TABLE OF CONTENTS

ARTICLE **PAGE**

FIGURE 4.3

MEDIATION AND ARBITRATION

Regardless of your best intentions and the best intentions of all parties in negotiations, not all issues can be resolved at the bargaining table (see Figure 4.4). Once an impasse is reached, there are usually some options to consider. Some states are right-to-work states that do not formally recognize labor organizations. Variations of this include states that prohibit contact with labor unions, states those that allow a nonbinding "meet and confer" relationship, and states that have established mediation and arbitration as means of resolving disputes and impasses.

In these cases there is an established procedure that usually begins with assignment to a mediator who uses various techniques to encourage the parties to work toward an agreement. **Mediation** is a nonbinding process designed to resolve a dispute between two opposing sides. The mediator functions more as a counselor to bring the groups together to resolve their differences. A good mediator can be effective with the right set of circumstances and if reasonable people are involved.

mediation
A process designed to resolve a dispute between two opposing sides

Fact finding is another method used to attempt to resolve disputes. Fact finding allows the parties to present their view of issues to an unrelated third party who then considers both sides of the dispute and makes a recommendation. Like mediation, fact finding is usually not binding. In contrast to mediation, in which the mediator works with both groups to find common ground and determine where agreement can be found, fact finding is an attempt to present what is known, that is, the facts, and

fact finding
Presentation of issues as parties view them to an unrelated third party who then considers both sides of the dispute and makes a recommendation

FIGURE 4.4 ◆ Good relationships between labor and management can minimize conflicts and improve negotiations.

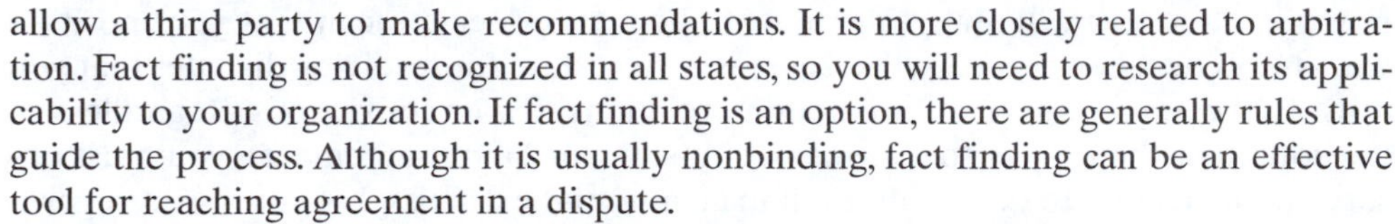

allow a third party to make recommendations. It is more closely related to arbitration. Fact finding is not recognized in all states, so you will need to research its applicability to your organization. If fact finding is an option, there are generally rules that guide the process. Although it is usually nonbinding, fact finding can be an effective tool for reaching agreement in a dispute.

If mediation and/or fact finding fails, there may be provisions for submitting the case to **binding arbitration**, in which both sides present their best last offer, and the assigned arbitrator hears the arguments and make a ruling. Although the rules can vary from state to state, generally the arbitrator is instructed to select one side or the other. When this happens, the arbitrator usually takes a "King Solomon" approach that allows each side to "win" an issue. In its purest form, arbitration encourages good-faith bargaining to try to get both sides to compromise. If either the labor or management side is way out of line, it risks losing a lot, since the arbitrator cannot negotiate and must select one side or the other.

binding arbitration
A process for settling a dispute between two parties in which both sides present their best last offer, and the assigned arbitrator hears the arguments and make a ruling

Although it is difficult to know all the rules of mediation, fact finding, and arbitration, you should familiarize yourself as much as possible and learn from your resources, the labor attorney, and the HR director. Ideally, resolution could be reached without the use of these tools, but as we all know, there will be differences of opinion, and it will be necessary to seek help to find a solution. The rules of each of these techniques can be specific to a particular state and may be confusing to those who do not regularly engage in these practices. Again, the recommendation is to take advantage of your network of experts. In addition, should you be required to be a part of any of these proceedings, anticipate that a lot of preparation will be required.

◆ STRIKES OR WORK STOPPAGES

The idea of a firefighter work stoppage, or strike, is a very distasteful thought. Fortunately, there have not been many recent cases of this type of job action, owing in part to the dedication of the people involved (firefighters are mostly in the work for its public service to the community), legal issues (laws banning work stoppages by public safety personnel), and the positive effect of negotiations that allow for mediation and arbitration. Yet, the threat of a strike remains in some areas of the country. Almost always there is a warning that this last-step action is coming. If you are unfortunate enough to be involved somehow, prepare yourself. Know what will be expected from you with respect to service continuity. You also will see your relationship with the firefighters greatly challenged. If you have any say so in the matter, do what you can to avoid this action. No one wins, and the effects last for a very long time.

◆ DISCIPLINE AND DUE PROCESS

One of the major roles of labor is protecting the rights of the individual. As a result, it is imperative that you know and understand the rules and regulations of your organization, any requirements of the labor agreement, and applicable laws. Regardless of your position on a particular issue, the most important thing is to follow due process, the formal proceedings carried out in accordance with the established rules. This does not mean that you will always agree with them, nor does it mean that your position will always be taken. It means that there is a procedure for resolving disputes that you must follow regardless of your emotions, relationships, or convictions. Regardless of the source of these rules (department, city, state, or labor agreement), you must observe them. Failure to do so will result in an automatic dismissal of your issue, regardless of the merits of your case (or your side of the story).

Other than negotiations, the issue of discipline of bargaining unit employees is the most likely area of contact between labor and management. You should have a good policy in place that establishes procedures and discipline options. These can be in your department rules (or policies or procedures) or in the labor agreement, or both. The policy should include a reaffirmation of management rights, causes for disciplinary action, and the employee's rights. Figure 4.5 is a sample document.

Any disciplinary action other than an oral reprimand or oral counseling needs to be documented. Some individuals recommend documenting *all* disciplinary actions. If this is the course you decide to take, documentation of an oral reprimand should state only that this particular action was taken, and details will be minimal. Disciplinary letters should be simple and to the point. They should include the facts of the incident; the parties involved; the specific violations of the rules, policies, procedures, or contract that were violated; and the corrective action taken or to be taken. Also a statement should be included regarding future violations not only of the specific rule being referenced but other potential issues. Members should not think they are allowed to violate every rule at least once! Lastly,

SAMPLE DISCIPLINE AND DISCHARGE POLICY

SECTION A. The City shall retain the sole right to establish, change, amend, and enforce rules for employees to follow, the right to warn, reprimand, lay off, discharge, demote, or transfer any and all employees who violate these rules.

SECTION B. After completion of the probationary period, no employee shall be suspended or discharged without cause.

SECTION C. Cause for disciplinary action shall include, but is not limited to: failure to observe rules of conduct established by the City; inefficiency or inability to perform assigned duties, excessive absenteeism, tardiness, failure to take a medical examination; dishonesty, or theft; insubordination; overt discourtesy to supervisors, visitors, or other City employees; failure to work with supervisors and fellow employees in an acceptable manner; gross neglect of duty; intoxication; use of alcohol or drugs on City premises or during working hours; failure to observe work rules (including rules in regard to dress and appearance); falsification of employment application or other records; or assumption of supervisory authority of advising or directing employees to disregard the orders of supervision.

SECTION D. In the event an employee is suspended or discharged, the employee will be entitled to the presence of a Union representative, if the employee so requests. The employee may file a grievance at Step Two of the Grievance Procedure. The role of the Union Representative shall be that of an observer.

SECTION E. The employee shall be furnished a copy of any new entry or disciplinary action and shall be given the opportunity to initial or sign such entry prior to its introduction into his file.

FIGURE 4.5

note that future violations will result in further discipline up to, and including, discharge.

SAMPLE DISCIPLINE LETTER

Dear Firefighter,

On January 10 you reported to work 30 minutes late. This is in violation of section 3 of the labor agreement and the Department Rules and Regulations. This is your second violation in the past 12 months. As a result you are suspended for 3 days. Future instances of this or other violations will result in further discipline up to, and including, discharge.

Sincerely,

You may also wish to have the offender sign the letter to acknowledge receipt. The statement should be worded so that it does not prohibit the filing of a grievance if there is a disagreement on the violation or discipline taken. The purpose of the signature is to verify receipt of the letter.

GRIEVANCES

In spite of everyone's best efforts, sometimes there will be disagreement about the handling of department issues affecting labor. Although it is hoped that issues can be resolved, differences in personalities and perspectives create differences of opinion, stubbornness, and even honest disagreements in interpretation of the labor agreement. When that happens, there must be a system in place to allow appeal of an action or decision, that is, a grievance procedure, to resolve these differences so the department can move on. A **grievance** is a cause for a complaint. The procedure is a method for attempting to resolve the complaint. View it as your own court system for dealing with situations that don't always work the way they should. Of course, there is a downside. Some labor groups have used the grievance procedure to bog down the fire chief and management. Be conscious of this and seek assistance if it is becoming an issue. The grievance procedure should include the steps of submission, appeals to higher authorities, and ultimately (as a last step) arbitration.

grievance
A cause for a complaint

How you handle yourself throughout the grievance process is important, almost as important as being right. When a grievance is submitted to you, do your best to be professional. This is not always easy. It is easy to say to someone, "do not take grievances personally." However, it is different when you are on the receiving end. This does get better with experience, but try your best to remain unemotional and understand that this is just a process.

Your organization should have a grievance form. This is usually provided by the union, as they should be the control point for filing a grievance. The form should include the obvious details such as name, rank, assignment, and the date of the infraction. It should also include the name of the person responsible for the alleged violation as well as citation of the section of the labor agreement pertaining to the violation and the facts of the case. Finally, there should be a place to suggest a remedy.

SAMPLE GRIEVANCE FORM

Grievant ______________________________

Rank ______________________

Assignment ____________________________________

Date of alleged* violation __________

Location of alleged* violation ___________________

Section of contract pertaining to alleged* violation _____________

Facts pertaining to grievance _________________________________

Suggested correction ___________________________________

Signatures of grievant and union representative

Steps of the grievance process

*Note—a form submitted by the union will not contain the word *alleged*.

After you receive the grievance you are expected to respond, usually within a given, preestablished time period. This is not an emergency and you have time to formulate your response. There are a couple of things to consider. Let your boss know. Remember, no surprises. Determine your boss's opinion on the subject. There is no point in carrying the issue forward if it will only get reversed at the next step in the grievance. Remember to be completely honest and forthright with your boss. You will not like the response if you paint one picture initially and other facts come out during an appeal. Tell it like it is and look for advice.

Besides obtaining the advice of your boss, also consult people with experience in this area, namely the labor attorney (not just any attorney, but one with special knowledge of labor law) and the HR director. Their expertise has previously been discussed. Based on the information gained from all these sources, prepare your response. Have one of your advisers (e.g., attorney, HR director) review your response. If it is acceptable, hold a meeting to deliver your response. Take good notes in the meeting because any additional comments made during that meeting may be brought up in future steps of the process. Figure 4.6 is a sample grievance procedure.

These suggestions are applicable up to the mediation and/or arbitration steps of the process. At that point, it is suggested that your labor attorney handle the process. Remember, his or her expertise is extremely valuable. He or she will handle the process. Your task is to be as well prepared as possible. Treat this as you would a court appearance. It is very similar. Be prepared by studying the case, the facts, your rules, and anything else of relevance. As you go through the process, be honest, remain professional, and control your emotions. It is a good idea to practice in advance. Since you do not do this regularly, training helps. As the chief you are expected to maintain your composure, exhibit fairness, and be the expert on this issue. This is another instance where how you do something is as important as what you do.

SAMPLE GRIEVANCE PROCEDURE

SECTION A. A grievance is defined as an alleged violation of a specific Article and Section of this Agreement.

SECTION B. If the grievance involves the discharge or discipline of an employee for cause, it shall be processed in accordance with the provisions of the City of Farmington Hills' Municipal Code and the Fire Department Rules and Regulations.

SECTION C. If any grievance arises during the term of this Agreement, such grievance (except those excluded in Section B) may be submitted to the following Grievance Procedure:

Step One. If an employee feels he has a grievance, he shall, within five (5) working days of the time the alleged violation occurred, present the grievance orally to his immediate supervisor or other designated supervisor as the case may be. Unless the supervisor determines otherwise, the meeting will occur immediately before the end of the employee's work shift. The employee's Union representative may be in attendance if the employee so requests. The supervisor shall submit his answer within three (3) working days after its presentation. If the grievance is not satisfactorily adjusted, the employee may submit a written grievance at Step Two.

Step Two. If the grievance is not resolved in Step One, the employee may reduce his grievance to writing on a grievance form provided by the Union and present the grievance to the Fire Chief, or his designated representative, for a written answer. The written grievance shall be filed within five (5) working days of the Step One answer. It shall name the employee(s) involved, shall state the facts giving rise to the grievance, shall identify the provisions of this Agreement alleged to be violated by appropriate reference, shall state the contention of the employee and of the Union with respect to these provisions, shall indicate the relief requested, and shall be signed by the employee. The Fire Chief, or his designated representative, shall give the employee an answer in writing no later than ten (10) working days after receipt of the written grievance.

Step Three. If the grievance is not resolved in Step Two, the Union may, within five (5) working days after the receipt of the answer in Step Two, appeal the grievance to the City Manager. The appeal shall be in writing and it shall include the written grievance and the Fire Chief's answer and shall specify the basis of the appeal. A copy of the appeal shall be sent to the Fire Chief. The Union may, at the same time the written appeal is filed, submit a written request to the City Manager for a meeting between the Union and the City Manager, or his designated representative, to attempt to resolve the grievance. The meeting will be at a mutually agreeable time and will take place within ten (10) working days after receipt of the written appeal and the request for a meeting. The City Manager, or his designated representative, shall give the Union an answer in writing no later than ten (10) working days after receipt of the written appeal. Additional time may be allowed by mutual written agreement of the City and the Union.

SECTION D. All grievances must be filed in writing within eight (8) working days from the time the alleged violation was to have occurred or they will be deemed waived. Any grievance not filed within the prescribed time limit or not advanced to the next Step by the employee or the Union within the time limit in that Step, shall be deemed abandoned. If the City does not answer a grievance within the time limits prescribed in this Article, the grievance will be considered automatically referred to the next Step of the Grievance Procedure. Time limits may be extended by the City and Union in writing; then the new date shall prevail.

SECTION E. Workdays for purposes of this Article, shall be Monday, Tuesday, Wednesday, Thursday and Friday, excluding observed holidays.

SECTION F. Any agreement reached between management and Union representative(s) is binding on all employees affected and cannot be changed by any individual.

SECTION G. A matter involving several officers and the same question may be submitted by the Union as a policy grievance and entered directly at the Second Step of the Grievance Procedure. Separate grievances, timely filed under the Grievance Procedure, arising out of the same or similar set of facts or incident shall be consolidated and handled as one grievance.

(*continued*)

SECTION H. If the grievance is not resolved at Step Three of the Grievance Procedure, and if it involves an alleged violation of a specific Article and Section of the Agreement, the Union may submit the grievance to the American Arbitration Association with written notice delivered to the City Manager within ten (10) working days after receipt of the City Manager's answer in Step Three, or, the day such answer was due. If no such notice is given within the prescribed period, the City's last answer shall be final and binding on the Union, the employee, or employees involved, and the City.

SECTION I. It shall be the function of the Arbitrator, and he shall be empowered, except as his powers are limited below, after proper hearing, to make a decision in cases of alleged violation of the specific Articles and Sections of this Agreement.

1. He shall have no power to add to, subtract from, disregard, alter, or modify any of the terms of this Agreement.
2. He shall have no power to establish salary scales or change any salary.

If either party disputes the arbitrability of any grievance under the terms of this Agreement, the Arbitrator shall first determine the question of arbitrability. In the event that a case is appealed to an Arbitrator on which he has no power to rule, it shall be referred back to the parties without decision or recommendation on its merits.

If the Arbitrator's decision is within the scope of his authority as set forth above, it shall be final and binding on the Union, its members, the employee or employees involved, and the City. The fees and expenses of the Arbitrator shall be shared equally by the City and the Union. All other expenses shall be borne by the party incurring them.

Claims for Back Pay. All grievances must be filed in writing within five (5) days from the time the alleged violation was to have occurred. The City shall not be required to pay back wages for more than five (5) days prior to the date a written grievance is filed.

All claims for back wages shall be limited to the amount of wages that the employee would otherwise have earned, less any compensation that he may have received from any source during the period of the back pay. No decision in any one case shall require a retroactive wage adjustment in any other case.

SECTION J. At the time of the Arbitration Hearing, both the City and the Union shall have the right to call any employee as a witness and to examine and cross-examine witnesses. Each party shall be responsible for the expenses of the witnesses that they may call. Upon request of either the City or the Union, or the Arbitrator, a transcript of the Hearing shall be made and furnished the Arbitrator with the City and the Union having an opportunity to purchase their own copy. At the close of the Hearing, the Arbitrator shall afford the City and the Union a reasonable opportunity to furnish Briefs.

The Arbitrator will render his decision within thirty (30) days from the date the Hearing is closed or the date the parties submit their Briefs, whichever date is later.

SECTION K. In cases involving disciplinary action, and where the disciplinary action results in a suspension greater than four (4) days or in discharge, the employee may submit the grievance to binding arbitration as set forth in this Article.

FIGURE 4.6

◆ DEVELOPING COOPERATION, NOT CONFRONTATION

Involving the union in the department decision-making process can be beneficial in building cooperation and an effective team. Although union representatives cannot, nor should they, be involved in everything, there are situations in which they can add value not only to the immediate decision but to the overall trust within the organization. Include a union representative on various committees. They have a vested interest in almost everything that is done. Provide the parameters (cost, time, personnel, etc.) and then allow them the authority to contribute. There are countless examples of great things that have happened with this level of cooperation.

One of the challenges a chief can face is that in many cases, especially in smaller departments, the entire department except the chief is a member of the bargaining unit. This sometimes challenges the loyalties of the senior officers, who rely on the union for representation yet are part of the management team of the department. Being a supervisor, manager, or administrator may require unpopular decisions. You must be explicit in all communications as to your expectations while acknowledging this potential conflict of interest. These senior officers may spend more time with the rank and file, and these relationships can, and most likely will, affect performance.

Good labor relations are very important to the effective and efficient management of a fire department even if there is no formal labor organization. The ability of labor and management to get along and work toward the common good of providing outstanding service to the community can determine the success of not only the department but the fire chief. The following are a few things to consider:

- Trust is essential.
- Emotional issues may be more important than the rational ones.
- Perception is reality based on an individual's personal beliefs and positions.
- It is important to communicate, communicate, communicate.
- Conflicts will happen and need to be solved, not viewed as personal in nature.
- Know the rules of negotiations and know your desired outcome.
- Labor and management must always remember why their jobs exist.
- Management must commit to good relations, not just talk about them.
- Mutual respect is essential.
- Knowledge of the roles of labor and management is important.
- Labor representatives should be included whenever possible when they have a stake in the outcome.
- Although important, negotiation is only one part of overall labor—management relations.
- Constant effort is required.
- Disagreements should be kept professional and impersonal; small disagreements should not be allowed to affect overall relations.

◆ SUMMARY

The importance of good labor–management relations cannot be overstated. The people of great organizations work together. You would be hard-pressed to find any great fire department (or other great organization in government or business) that has constant conflict and confrontation. The days of hard-line labor relations are no longer effective in today's environment. Cooperation is much more desirable and produces better results. Commit the energy needed, learn as much as you can, and always look for ways to build your relationships.

ACTIVITIES

1. Review the labor agreement in your community. What changes would you suggest from a management perspective?
2. Research the arbitration law for your state, if one exists. Summarize its applicability.
3. Review the disciplinary process for your community. Suggest changes to the process that would make it more applicable to your department.
4. Review a grievance submitted to your department. Determine the steps that were taken to resolve it. Would you change or handle anything differently based on the outcome?
5. Create a list of issues in which the fire chief and union president have a vested interest. Determine which present the same goals to both, that is, which ones are mutually beneficial. Identify methods for coordinating efforts to reach conclusions beneficial to both labor and management.

◆ CASE STUDY

CONTRACT NEGOTIATIONS

You are involved in contract negotiations with the firefighters union. Although you do not generally get involved in the financial aspects of the process, you are expected to provide input on the noneconomic issues that affect your operation. During this round of negotiations, the exchanges are especially spirited in comparison with previous years' sessions. The lead negotiator for the city is in the position for the first time and has not established a good relationship with the negotiating team for the firefighters. From your experience, you believe that the negotiations are not going well even though the firefighters are not being unreasonable in your opinion. You believe that the reason is not the issues but the personalities. The change in the city's team has not helped. Those who had been involved in the previous negotiations decided to pass the responsibility to a new team and are not interested in intervening. You are concerned that the goodwill previously established will be greatly diminished.

What would you do?

Change Management

KEY TERMS

communication, p. 55
culture, p. 55
leadership, p. 55
MBWA (managing by wandering around), p. 56
motivate, p. 55
trust, p. 55

The world is changing more rapidly than ever before and doubtless will continue at this pace, and the fire service will follow suit. The fire service has sometimes been disparaged by insiders with the comment "200 years of service unimpeded by progress" (or some variation thereof). Although this opinion has been expressed, there is also a realization that there has been tremendous change in the fire service, particularly in the last 30 years. You as the fire chief must embrace and prepare yourself and the organization for the inevitable change. The preparation, in most cases, is more important than the change that will follow.

◆ REASONS FOR CHANGE

You must understand the reasons for change. Change will come either by crisis, by outside direction, or from inside the organization, usually through you. There are many examples of changes that occurred in response to crisis. For example, many of the fire safety codes are the result of disasters. Large fires or fire deaths create pressures to operate differently so history is not repeated. Outside direction can come from your boss or the policy makers in your community (who may even be politically motivated). Rapid growth in a community or a significant decline in revenue may result in new or closed fire stations.

Other changes are the result of industry changes in laws, standards, and regulations. Occupational Safety and Health Administration (OSHA) regulations such as two in/two out have an impact on fire departments. Departments must adjust and adapt to OSHA regulations or face potential sanctions. Regardless of your position on the rules, compliance is necessary. Federal guidelines such as the National Incident Management System (NIMS) and standards such as NFPA 1500, "Standard on Fire

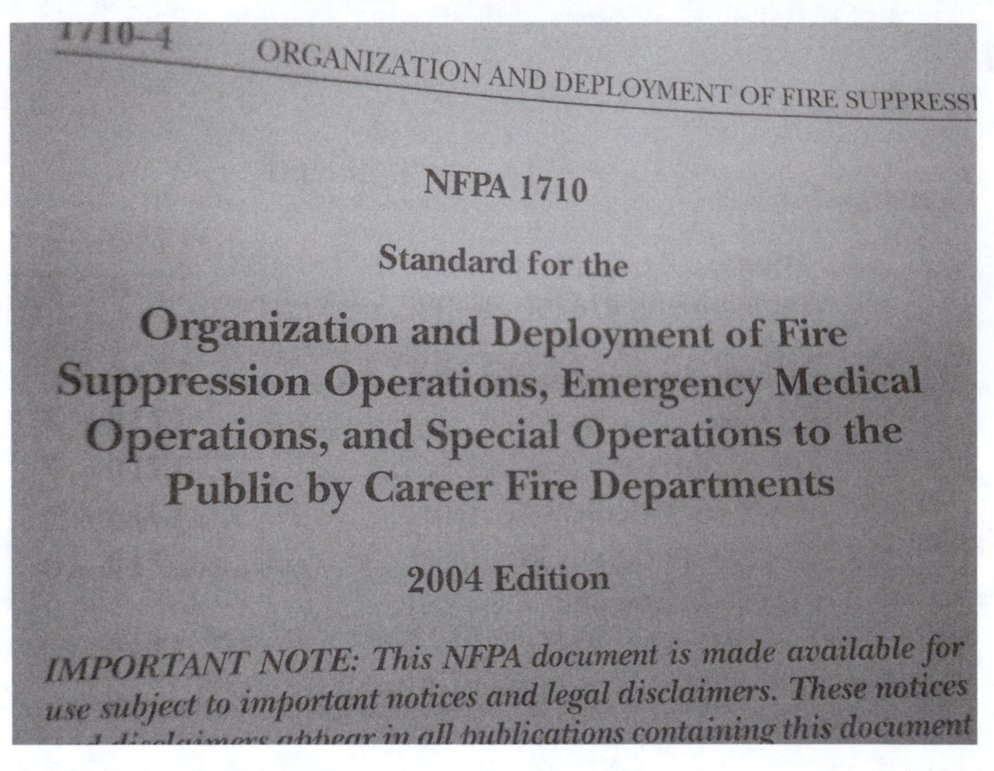

1710–4 ORGANIZATION AND DEPLOYMENT OF FIRE SUPPRESSI

NFPA 1710

Standard for the

Organization and Deployment of Fire Suppression Operations, Emergency Medical Operations, and Special Operations to the Public by Career Fire Departments

2004 Edition

IMPORTANT NOTE: This NFPA document is made available for use subject to important notices and legal disclaimers. These notices

FIGURE 5.1 ◆ NFPA 1710.

Department Occupational Safety and Health Program," and NFPA1710, "Standard for the Organization and Deployment of Fire Suppression Operations, Emergency Medical Operations, and Special Operations to the Public by Career Fire Departments," (see Figure 5.1) and NFPA1720, "Standard for the Organization and Deployment of Fire Suppression Operations, Emergency Medical Operations, and Special Operations to the Public by Volunteer Fire Departments," and voluntary programs such as the accreditation process from the Center for Public Safety Excellence (CPSE) should prompt an evaluation of the organization and, it is hoped, result in positive change. Although standards such as those from the NFPA are not law, they may have the force of law during litigation if no other standards or laws were used as a basis for decision making. Laws mandating levels of licensure and certification for firefighting and EMS compel the implementation of new and different training programs. One of the challenges that you face is to know the regulations that affect your operation. Of course, there are many, and it is difficult to keep up. Regardless, you must do your best and should be able to list the various laws, standards, and regulations that have affected your operations and the steps you have taken to comply.

If you compare the fire service of 30 years ago with today's, you can list some very significant changes that have occurred. Some of them have been embraced, but many have been accepted grudgingly. Some of the lessons to be learned are that many things will happen, with or without your approval. Look at the following list and think of how things today are different and recall whether there was immediate acceptance:

- Diversity—more women, people of color, different religious backgrounds (see Figure 5.2)
- The skill set of entry-level employees

FIGURE 5.2 ◆ Women firefighters are relatively new to the fire service.

FIGURE 5.3 ◆ Computers have changed and will continue to change the way things are done in the fire service.

- EMS—BLS, ALS, transport
- Run volume—fire calls down, EMS up
- Protective clothing—SCBA, turnout equipment
- Vehicle safety and driving standards
- Incident management systems (Incident Command System)
- The National Fire Academy
- Respiratory standards
- Technology—all the things related to computers (see Figure 5.3)
- Job functionality—career path

You should be able to add to this list not only general but specific changes that have taken place in your organization. Consider these questions: Are you still doing business the same way? Do you have the same number of people, the same apparatus, or equipment? Are the expectations of the public the same? Certainly, you can offer an opinion on these questions. There have been changes in EMS, response to incidents involving hazardous materials, response to and preparation for potential acts of terrorism, and involvement in emergency management and disaster planning, to name a few. No doubt you can add to this list.

Some of the changes are out of our control (to some extent) such as run volume, others may be imposed on us by the legal system (from federal statutes such as the NIMS, ADA, or FMLA), others through tragedy (emergency vehicle safety), and some through visionaries who took an idea to implementation (the National Fire Academy). Regardless, you should get the picture that in spite of our strong attachment to the traditions of the fire service (good and bad), we really have moved the profession forward to adjust to the ever-changing world in which we live.

◆ CHANGE CULTURE AND LEADERSHIP

Change is inevitable. Those that embrace it and work with it will succeed, not only immediately but also in the long term. There is value to change. This is an important concept not only to the fire chief but to the entire department. It is the responsibility of the fire chief to create a **culture**, a particular set of attitudes that characterizes the fire department, that allows the individuals and organization to recognize, accept, and even suggest change to move the department forward. This culture helps **motivate** all the employees, that is, creates enthusiasm in, interest in, and commitment to accepting the change and prepares everyone for something that will happen with or without their approval!

culture
A particular set of attitudes that characterizes the fire department

motivate
To create enthusiasm, interest, and commitment

One of the key components of leadership is dealing with change. It has been said that "*managers* do things right and *leaders* do the right thing." This is true about the role of leadership in change. **Leadership** is the ability to guide, direct, or influence people, and people need help with change. It is your task as the chief to convince personnel that change is inevitable. It is also your job to recognize what will change as a result of outside forces and what needs to change because of your initiative to continue to do the things necessary to be successful. Change involves risk taking, which is a big part of being a leader and also part of the personality of firefighters. Not all people, including chiefs, are leaders. Often, people are promoted based on their technical prowess, expert knowledge, or salesmanship, all of which can be important. Leadership abilities and qualities are hard to measure quantitatively, so promotions and appointments don't always consider this important role. Yet, in order to be successful and sustain the good services provided and improve on or add to those that need it, the fire chief needs to be a strong leader. The fire chief must create an organization that embraces change, accepts change, and is continually prepared for change, and the chief must be prepared to lead.

leadership
The ability to guide, direct, or influence people

◆ COMMUNICATION AND TRUST

There are two major factors in change—communication and trust. You cannot create an organization willing to change without good communication throughout the entire department. **Communication** is the exchange of information by means of speaking, writing, listening, and even through unspoken signs or behaviors. You cannot build trust without communication.

communication
The exchange of information by means of speaking, writing, listening, and even unspoken signs or behaviors

How do you communicate? Is there a system in place that allows two-way communication? This is done in a variety of ways: through meetings (formal and informal), in writing (newsletters, memos, notes, etc.), verbally (official and unofficial), electronically (email, cell phones, and pagers), nonverbally (through your actions), and other proven methods.

As mentioned previously, great communication is an essential element of building trust. **Trust** is confidence that everyone will be treated fairly, truthfully, honestly, and with respect. There should be no surprises and no unknowns. This is important for the

trust
Confidence that everyone will be treated fairly, truthfully, honestly, and with respect

simple reason of keeping people informed. People have a natural fear of the unknown. The more information you can provide them, the more likely they will be comfortable. You also keep them engaged in the process, which helps gain support and solicit suggestions for improvement. People involved in the change process will work much harder to ensure the success of the change than those who are merely observers. If their ideas contribute to the change, they have a vested interest in seeing that it is successful.

Obviously there are some challenges to good communications, or every organization would have all the bases covered. Here are a few examples:

- Managers may find communication difficult because they have not been prepared. They may need to improve their oral or written skills. The obvious solution is more training and education. Experience will also help in this area.
- Managers may know what they want or need and where they want the organization to go, so they think everyone knows. Either intentionally or unintentionally, the result is that key information is not shared.
- Some things seem so obvious to the people in charge. As chief you have above-average skills (or you would not have been promoted). Remember that everyone in the organization does not move at your pace.
- The fire chief assumes that others know or will ask if they don't. People are sometimes hesitant to talk to the "boss," or they may not want to appear ignorant on a subject.
- Because of the human instinct to resist change, some communication starts from a "no" position, which slows down the process. Some people may like things just the way they are, so they really don't want new "stuff" adding to an already full and complex workload.
- Problems can be complex, but in today's world everyone wants answers in 30-second sound bites. There is instant access to information and solutions are perceived as simple, so people can't understand why things take so long or why things don't happen like they are portrayed on TV! People are not always willing to read lengthy explanations. They want summaries.
- The diverse workforce responds differently, sometimes requiring different forms of communications to be effective. For example, today's generation has grown up with technology and is very comfortable with it and may prefer the newer ways, whereas some of the senior employees may not share that same comfort zone with text messaging and other technological innovations.
- There also can be a natural skepticism between boss and worker, so that no matter what is said, it won't be believed until proven.

SUGGESTIONS FOR IMPROVING COMMUNICATIONS

Challenges to good communications do not mean that the effort should not be made. There are ways to overcome the challenges and make things better. This does not mean that all problems can be solved, but their impact can be lessened. It is up to the fire chief to recognize the challenges and seek out solutions. A failure to do so will make change extremely difficult if not impossible. Here are a few suggestions:

MBWA
(managing by wandering around)

- **MBWA or managing by wandering (walking) around**. MBWA means asking and observing, not waiting for information (see Figure 5.4). If you stay in your office, you do not learn what is happening. Even if you have an open-door policy, most people won't use it because you are the "boss." Even if you do get information, it will be presented through the "filter" of the person presenting it.
- In spite of the previous bullet implying that it may not work, have an open-door policy. Although most will not use it, you will gain some information, especially when you do this in addition to MBWA. It also sends an important signal that you can be approached anytime,

FIGURE 5.4 ◆ The fire chief needs to get out of the office and talk to the membership wherever the opportunity may present itself.

though few will have the confidence to do so. There are some precautions, however. Do not let people circumvent the chain of command when they need to follow it, and do not do anything that might be interpreted as undermining your staff or officers. You can address this potential problem by always leaving the option open to refer the individual back to their immediate supervisor.

- Conduct many meetings. These can be of all types—formal, informal, planned, impromptu, large, or small. Those in attendance should include any and all who will be affected (to the extent possible) and those responsible for any subsequent action. Of course, there is a downside. Although meetings are very important, they can be detrimental if they affect productivity and waste time. Find the balance between necessary meetings and time wasters. Also, it is helpful to have agendas for meetings so you stick to the business at hand. Although it may seem contrary to the preceding, meet only when it is necessary. Sometimes you can "meet" yourself out of good communications.
- Utilize a newsletter and memos to get factual information in the hands of those who need it. Do your best to get this information out before the rumor mill gets started.
- Rumor control is important. In all organizations, there is an element of communication through rumor. It is combated with good, factual, continual communication and a policy of no tolerance for rumors. Consider publishing a document that recognizes and responds to rumors.

How do you create a culture of change that allows for the implementation of new programs and ideas? Start with what your role is. Although fire chiefs are required to be both leaders and managers to be very successful, there are situations that require the leader in you to come to the front. Change management is one of those. You need to clearly exhibit leadership to your personnel when establishing your role in the change process. They will then be able to identify who and what you are and allow that acceptance to occur. Then you can start building their trust, day by day, with every individual.

Trust is built not only with good communications but by doing what you say you will do. People will follow leaders they trust. Be assertive and aggressive. Meet your established deadlines and exhibit a sense of urgency when dealing with change or addressing problems or issues. You need to create comfort so you can ask questions without placing people on the defensive. On the flip side, allow others to question you. Can you handle tough questions without exerting the influence of your rank? Practice what you preach and walk your talk. In this area it means do what you say, don't make promises you can't keep, be consistent in your message, allow others an opportunity to express themselves without fear of retribution (good or bad), and embrace change yourself.

Be an expert, know your stuff, and be a good example. Continue to train and educate yourself. Read, attend conferences and workshops, consider higher education, and build your network. Although you don't need to know everything, as the fire chief you will be expected to have a good grasp of your profession, but in demonstrating it, control your ego. All chiefs have an ego or they wouldn't be chief. Keep it in check. Finally, in your role as the leader of change, include as many people as possible as often as possible. It is not possible to include everyone every time, but the more people involved, the more often the correct information gets out.

Although it is possible to create a change culture, human nature may make this more difficult. You must prepare the organization for change, not just implement it. This is very time consuming and takes a great deal of energy, especially in the early stages. If you are not prepared to commit the time and energy, do not expect good results. You also need to be flexible. It is nearly impossible to predict the future. Circumstances will change; plans will not be perfect. Be willing to change on the fly to adjust your plans and correct mistakes. If people observe you making adjustments, they will gain confidence that the final outcome will be as good as can be expected. A final suggestion is to minimize the rules wherever possible. Keep people focused on the task at hand and the results, not policies and procedures. Allow people the freedom to act without fear of consequences provided the motivation is proper. Change and innovation are never perfect, and there will be trials and tribulations along the way. Accept them. If you wait for perfection in a plan hoping to eliminate any potential problems, you will never take action. You minimize the risks but still need to move forward.

◆ TEAM BUILDING AND TRUST

Trust was discussed earlier, but it needs to be mentioned again because it is important throughout the organization. No matter how hard you try to build that trust, you are reliant on others—your staff and officers—to do the same. This issue increases as the size of the department grows. Realistically, you can supervise only a limited number of people. You rely on others to do their job, part of which involves this element of trust. You need to support others, that is, support the individual while dealing with the outcomes of the issue. You can delegate authority over things, but you can't give up the responsibility. Stand behind the choices that your employees make when they are acting on your behalf and within the scope of the direction you provided. Correct wrong actions by accepting responsibility, and make accountability for actions the norm. Avoid blame (unless it is obvious sabotage or the lack of a sincere or honest effort). The willingness of people to change is based on mutual trust, a two-way street.

Avoid micromanaging. Set expectations and make clear what you want the end result to look like, not how to get there. To get trust, you must give trust. This does not mean that you cannot ask questions. You need to know what is going on and why things are being done. Interest in a project or action is not necessarily micromanaging. You need to allow others to do their job. You decentralize the decision making. Delegate as much as possible and monitor the results. This helps prepare subordinates for future opportunities to step up as a chief, if and when needed. You help and mentor the process; you do not take over or always dictate the change. You will be pleasantly surprised at the quality of decisions made if you have good people, provide clear direction, give them the tools, and let them go. For example, consider the purchase of fire apparatus. As chief, you probably never ride on the apparatus. Why do you need to dictate the type of apparatus or features on the piece? Select the right persons for the job, provide a budget, tell them the expectations of performance (types of calls expected to handle), give them access to information, and turn them loose. You will get a great piece of apparatus that is supported by the membership. They will take great care of it and make it work.

◆ CULTURE FOR POSITIVE CHANGE

The following are some practical things you can do to help create a culture for positive change:

- Start small. Easy-to-attain and acceptable projects build support for future endeavors.
- Promote early successes. Nothing breeds success like success.
- Do not do token things because they are easy. Do things that are needed and credible.
- Do not start with issues that may be threatening in any way. For example, although some work rules may need changing, they are not the best place to start. They affect your employees, so any proposed change has the potential to be misinterpreted and not accepted.
- Allow choices if appropriate. If someone has a different idea and it works, try it. It doesn't always have to be your idea to be successful.
- Although a sense of urgency is encouraged, you cannot change faster than the organization will accept the change. You do not want to get too far ahead of the department.
- Be conscious of labor–management relations. Labor representatives will want to be involved. Do not view them as obstructionists but partners with a vested interest. This applies whether or not you have a formal, recognized union.
- Know the power brokers regardless of their positions. Although there are very few people who can promote a successful project, there are many who can stop or slow down a process with very little energy. There are informal leaders in any organization, and they must be considered as part of any change process.
- When you are ready to go, be impatient. When the groundwork is in place, "pull the trigger." Demand change so you can get instant results.
- Be consistent with change (a seeming oxymoron!).
- Do the right thing. Don't "do the thing right."

As you begin the change process, recognize the steps needed to improve. There is so much that can be done, but you need to be selective based on the needs of your organization, the resources available, your personnel, and your background and instincts. Don't change just to be different (for change's sake) but to get better. Also remember that much change does not require a great deal of financial investment. Do what you can afford.

Creating a change culture in an organization is as much about the individuals as it is about the group. You should understand the relationship and the need to consider individual issues. Regardless of what happens collectively, much of what is done is the result of a single person. You cannot disregard this while you are trying to create the culture of change in the fire department. Key individuals really do drive the department, and their action or inaction can make or break your programs.

The preceding discussion was about creating organizations that are not afraid of change and actually consider it normal. These organizations embrace change, which makes it easier consistently to pursue new and improved ways to provide existing services and add more that benefit the community. Of course, there is the need to change the culture of an organization regarding specific issues and items. For example, the "smoke eater" culture had to be changed so that firefighters would use self-contained breathing apparatus (SCBA) all the time. For many generations the image of a firefighter was one of an individual who didn't need the extra protection. It has taken quite some time, most likely a generation, to change the attitude and culture regarding the use of SCBA. There are still individuals and departments that do not use SCBA all the time, but the practice is changing. The change to using SCBA occurred because of training, policies, and enforcement with some added education regarding the health hazards.

These elements are a part of any change in culture on a specific issue or item. Sound policies, to be in place, they need to be enforced, personnel must be trained, and members must be educated (or sold) on the change so that they believe it is beneficial to them and the organization. There have been shifts in many parts of the fire service including emergency medical service, special response, fire prevention, and health and safety. All the changes required a new way of looking at these things. None of them changed overnight and required a plan and continual diligence.

◆ MOTIVATION

Individuals are motivated by their reasons not yours. To motivate, for these purposes, means to give someone the incentive or reason to do something. They do things that they see as beneficial and logical to them. You can create an atmosphere that encourages innovation and change but ultimately the person makes the choice to act or not act. People are motivated for two basic reasons: fear or passion. The fear can be a positive or negative force. Some people have a fear of failure in whatever they do, so they try very hard to be successful. It is part of their competitive nature. A negative force is fear of punishment if something does or does not happen. The penal system is designed to motivate people to behave within the guidelines of society or else they will be punished. The fear of reprisal does not generally work for the long run. It is the individual's perception of fear that creates the positive motivation, not your wielding a big stick.

Challenging an individual can motivate a person if it is handled correctly. Sometimes, telling people they cannot do something results in their efforts to prove you wrong. As long as the final outcome is what is desired, this can be good. However, sometimes a challenge can backfire if you unintentionally offend someone. For example, a chief might say, "If you don't like the way things are, you can find another job." Although some people might react the way the comment was intended, to motivate, others might respond aggressively. There may be times to use this approach as a motivator, but you should use extreme caution in how you turn people down or tell them you don't want something done in a particular manner. You do not need to be an amateur psychologist. Do the right thing—if it results in motivating the person, all the better.

The other motivating factor is passion. People do things because they are passionate about them. Some say that money motivates. But many studies have shown that money does not elevate people to higher achievement. Employees may be upset or dissatisfied if they are not adequately compensated (in their perception), but they will not overachieve for money alone. Consider professional athletes. The superstars make huge amounts of money, but larger contracts do not inspire them to greater achievements. If they are unhappy with their compensation, they will seek other opportunities to get what they think they deserve. More money does not make them run faster, throw harder, or jump higher.

Given a choice, people will pursue activities and achievements based on their passion. Consider hobbies. Many people commit much time and energy in their spare time pursuing something they really enjoy. It is the passion for the work that keeps them going. Obviously, they do not get financial compensation for their hobby. If you understand this concept, then you can reasonably deduce that those that are passionate about their work will be the high achievers.

Although motivation may be classified as described above, there is a subset of motivation that is applicable to firefighters, that is, recognition. Many firefighters have type A personalities; they like formal and informal recognition. This may be tied to their desire to make a difference in their communities through service to others that may involve saving a life. This recognition can be as strong as passion. You need to understand the role of recognition in keeping personnel focused on the mission of the department. Always say thank you. Send notes and letters of appreciation. Consider a formal awards program. Most recognition can be simple and inexpensive. Do your best to take the time to provide the recognition that is deserved.

◆ POTENTIAL BARRIERS TO CHANGE

Not everyone is ready for change. Those people have their reasons and you need to accept them and develop a strategy to work with them. Also, you need to realize that not everyone will be a shining star. Despite the fact that no one looks at him- or herself as an average performer, there are C students in your organization. They also have various strengths and weaknesses, meaning they can help in some areas and might be best kept out of others. Some factors to consider when dealing with individuals who are averse to change are listed here:

- They may be near retirement and do not want things different.
- They are comfortable with the way things are done.
- They have other things outside your organization that are of more interest.
- They don't believe that the change will last. They perceive that you are always doing things different for the sake of doing things different, not for the results.
- They fear the unknown.
- They have it very good and do not want to risk a change.
- They deny the need to change; everything seems to be fine.
- They may be jealous or have uncontrolled egos.
- They may want change only on their terms.

- They don't need change—everyone around them does!
- They are skeptical of the potential benefit.
- They believe it can't be done or it can't be done that way or that fast.

Although the goal is the creation of an organizational culture that allows and even embraces change, it boils down to individuals and their willingness to accept change. There is no magic to working with individuals. The need for good communications and the development of trust were discussed. Collectively and individually all chiefs know the value. This remains very important in one-on-one dealings.

Individuals need to be prepared for change. One key component is to continually train and educate the employee. Firefighters who continue to grow through education see the benefit of change and better understand the big-picture perspective. Education is important not only for change but for the overall development of the firefighter, which leads to better service to the community. There needs to be general education and specific programs targeting change and its effects. Knowledge is good for both the individual and organization. It can help not only with the acceptance of change but in the generation of new ideas and suggestions for improvement. Further, ideas that firefighters initiate often gain more grassroots support.

◆ POLITICS

Sometimes change is needed and is logical. Why doesn't it happen? One word—politics! So much of what happens in the fire service is as much about politics as it is about standards and mandates. Politics was discussed and defined in Chapter 2. Please refer to that chapter for more detail, but some considerations apply here to change. Know what your boss wants and is willing to support. Do not go out on a limb without the understanding and support of the person you answer to. Get the buy-in early, select issues that can be supported, and use the power of the boss to help. Make sure your boss is on board. Do not risk an important issue by not knowing how your boss feels about something. If you attempt to make a significant change without the boss's knowledge, he or she may be forced to render a decision that is contrary to your efforts. Remember, in this area, surprises are not good!

Governments and bureaucracies can be a great barrier to change by the sheer size of their organizations and their rules. Think of a ship. The smaller it is, the easier it is to turn. Larger vessels take more time and space. The same can be said of change in organizations. The larger, more complex departments take more time and "space." Individual freedom to change may be discouraged and restricted. Other related departments or agencies may also set up roadblocks. Learn who and what can potentially create obstacles, then work toward solutions to the challenges presented. Know the system and have other people who know the system help you navigate through the bureaucracy.

◆ MIDLEVEL MANAGERS AND CHANGE

Midlevel managers are extremely important to creating a change culture. They are closest to the workers and carry the message from the top. They also can carry messages to the top as long as they feel comfortable doing it. Although there may be an open-door policy, people need to be willing to walk through it. Remember not to shoot the messenger. By nature of being the boss, you will be isolated, and many in your organization will choose not to interact with you. Anyone who is willing to do so is very valuable.

Midlevel managers have a huge influence on their subordinates. They may or may not directly report to the chief. If they do not, they will respond to their immediate boss, the one who does their evaluation. Midlevel managers are mostly interested in the people closest to them, and their work is controlled by their immediate boss. They are also closest to the firefighters, generally have their support, and more closely identify with them than with the chief. So, you have a challenge to make sure they are engaged in the process and supportive. Keep this in mind and commit the time and energy necessary.

◆ SUMMARY

There are no magical gimmicks that guarantee results. The fire chief's job is about time commitment, communications, and trust. Make sure you are getting results, not just looking and sounding good. Change involves outcomes and improvement. Without these, cynicism grows. If it doesn't make it rain, a rain dance is no good.

Remember your human relations and continually build them. People are the key to effective change. Be wary of the current buzzwords. Gimmicks do not last. Over the years you have heard the terms *reinvention, empowerment, downsize/rightsize, total quality, continuous improvement, benchmarking,* and others. These are just various ways of getting to the core values of organizations that are change-ready. There are no shortcuts. Use common sense. Change is an ongoing process that is time consuming and requires hard work. Continually review and evaluate and make changes as needed. Finally, institutionalize change that works, and more important, institutionalize a change culture in your organization.

ACTIVITIES

1. List the three most significant changes in your department in the last 10 years. What elements were most responsible for the success of the change?
2. Develop a strategy to implement a significant change in your organization. Identify something to change and include all the elements needed to be successful.
3. What are the major barriers to change in your organization and what would you do to overcome these barriers?
4. Discuss the cultural issues that affect change in your organization. Which parts support change and which ones make change more difficult?
5. List five elements of the fire service that are likely or need to change in the next 5 years. What can be done to help make those changes successful?

◆ CASE STUDY

TRANSITIONING TO A TRANSPORTING AGENCY (EMS)

Your department provides medical response with department paramedics (cross-trained firefighters); however, the patients are transported by a private ambulance company. In the typical incident the fire department arrives first and initiates treatment. The fire department paramedics complete their work and the patient is then delivered to the hospital by a private ambulance company with one department medic on board and the department apparatus following behind to pick up the personnel from the hospital. As is acceptable practice, the private ambulance is permitted to charge a fee for its transport service. The department receives no reimbursement for its efforts. You have determined that the best way to remedy this situation would be for the department to complete the transport.

Develop a strategy for implementing this change.

Emergency Response

CHAPTER

KEY TERMS

apparatus, p. 69
equipment, p. 69
performance, p. 65
professionalism, p. 67
proficiency, p. 73
strategy, p. 74
tactics, p. 74

Few fire officers ascend to the top spot without adequate preparation for taking command of an incident. It is clearly a major strength of senior officers, especially those who are promoted through the ranks. Further, there are numerous training opportunities with respect to tactics, strategy, and incident command. Likewise, emergency response is the "fun" part of the job, so individuals naturally commit the necessary time and energy to get better. This chapter is not intended to address those aspects of the job. The purpose is to view emergency response from the fire chief's perspective.

The importance of emergency response to the fire chief is not personal but personnel preparation—the preparation of the department members to respond. It is the bread and butter of the fire service and the reason for its existence. Though there are many ancillary duties for a fire department, emergency response is obviously the most important function by far. Most everything else is done to support the service provided when someone has an emergency (or perceived emergency) and calls 911.

The expectations for **performance**, or the effectiveness of the way your personnel do their job, during an emergency call are very high. In fact, those who rely on firefighters to help them when things have gone bad expect perfection or near perfection. No one wants you to send the "C" team; they want the best. Of course, they weren't anticipating an emergency so they haven't shopped around for the best. People in a crisis have no choice, so it is imperative for the responders to be ready for whatever situation might be presented (see Figure 6.1).

performance
The effectiveness of the way personnel do their job

FIGURE 6.1 ◆ Fire department personnel must always be ready to perform at their best, even if not called on to do so every day.

◆ MOTIVATING PERSONNEL

All personnel must be prepared to perform at the highest level in all potential emergency situations. The chief must continually communicate the importance of this preparation by supporting quality training of personnel and demanding high performance. A significant part of this task is motivating personnel to be the best they can be and promoting an organizational culture that pursues perfection in the delivery of service. Training must include not only the necessary skills but continued repetition and practice to ensure a high level of performance. Neither the chief nor the training division can deliver enough training to accomplish this goal. Individual members must accept their responsibility to pursue excellence. Training and the subsequent practice create actions that are second nature, so the firefighters and officers can concentrate on the emergency at hand, not on skill issues that should be automatic.

◆ CITIZEN EXPECTATIONS

Based on the content of numerous complimentary letters that fire chiefs receive, the public is impressed with three things that the fire service provides when emergencies occur: rapid response, kind and caring service, and professionalism. The following represents an actual letter received from a citizen. Note that the elements of response, caring, and professionalism are included.

> Dear Chief,
>
> I just wanted to inform you about how impressed I am with your staff. My husband made a nonemergency call on February 25th at 5:30 a.m. because our carbon monoxide detector went off. We live in the Main Street Apartments. The gentlemen from the fire department arrived minutes after our call. They were extremely courteous and friendly. I was also very impressed because one of the firefighters called us the next day to see how we were doing. (I can't remember his name, as he mainly spoke with my husband.) We really appreciated the call. It is so nice to know that the fire department truly cares about our well-being, unlike our apartment management. Keep up the good work! We feel a lot safer knowing that you guys are out there!
>
> Sincerely,
>
> Jane Doe
> Apartment 5
> Anywhere, USA

The lesson to be learned is to be mindful of what is important to the people who are responsible for the existence of the fire department. Everyone in your organization needs to understand these elements. Response time needs to be minimized on all calls, not by driving faster but through preparation and turnout time. It can also be improved in the dispatch center, though many fire departments today rely on centralized dispatch centers, and the chief has minimal control of the operation. Goals and policies corresponding to those goals need to be established with respect to turnout time. What is acceptable in your department? Do all your members know what it is and are they held to that standard? More important, do you and your senior officers hold everyone accountable to meet the established standard. Communicate the importance to everyone, reminding them that the public and taxpayers value a rapid response.

The second element identified by the public as a valued part of your response is kind and caring service. Be nice! You probably learned that from your mom. Even though everyone may know this is important, it is often not overtly communicated to department members. Typically there is little offered to firefighters during training to help them improve their customer service skills. What they know comes from their instincts and by observing others in the organization. Customer service training needs to be included at all levels of the fire department. Remember, this is important to those on the receiving end of service.

The last item listed in most letters from the public is **professionalism**. Although the differences between career and volunteer organizations can be discussed, here professionalism means the skill, competence, or character expected. Regardless of the makeup of your department—career, volunteer, combination, or other—you must emphasize the need to present your organization as very competent and professional. Two things contribute to the perception of the department: appearance and action Make sure all your members present themselves well and reflect your values. Dress and act appropriately. You know what looks good and makes a positive impression. Provide uniforms that reflect the image you want and make sure all your members

professionalism
The skill, competence, or character expected

maintain what has been issued to them. This also includes turnout gear. Looking like you know what you are doing is significant in gaining control of an emergency and sends a strong message. Communicate this often to your membership.

The second element of professionalism is competence, and this is accomplished through training and practice. In sports there is a saying that "you perform in the game like you practice." This is true in the emergency service. Your training programs need to be results-oriented with the target being outstanding performance during an emergency. Many of the emergencies are simple (and even routine if one dares use that word), but you will be judged on your performance during the really challenging times. Prepare for the tough jobs!

◆ THE ROLE OF THE FIRE CHIEF

The actions of the chief with respect to personnel, training, expectations, apparatus, and equipment set the tone for the performance of the department during emergencies. To prepare the department, the chief must understand his or her role; the resources available, including personnel; the emergencies requiring response (including daily routine incidents and those occurring less frequently); and the challenges presented that may make this preparation more difficult.

Just what is the role of the fire chief? As mentioned previously, it is not necessarily important to concentrate efforts on your own preparation for responding but on the department response (and even the community response). To prepare for response, you must know the emergencies likely to be encountered. There are routine, day-to-day responses that require a certain level of readiness, and larger, less frequent, but high-consequence incidents that require a risk assessment to determine needs. These incidents can include major fires, disasters, mass casualty incidents, hazardous material emergencies, special rescue situations, and even response to acts of terrorism. They vary from jurisdiction to jurisdiction. Not everyone is faced with the following: wildland–urban interface, high-rises, snowstorms, hurricanes, earthquakes, railroad emergencies, or industrial sites. You must prepare for the hazards you are most likely to encounter.

If money, time, and personnel were no object, it would be easy to prepare, but, of course, there are limitations. This is one thing that makes the job more challenging. Personnel expect to have all the resources needed including adequate staffing and equipment. Since there are no blank checks, priorities must be set that allow the fire department to meet the expectations of the community. The big picture requires an assessment of the responsibilities of the department, expectations, and understanding of what it takes to provide the various services, personnel, and resources. With this information, the fire chief must be able to influence the policy makers and those controlling the resources of the core needs of the department. The preparation for response can then be matched with expectations and finances.

◆ APPARATUS AND EQUIPMENT

What about apparatus and equipment needs? **Apparatus** is the "rolling stock," that is, vehicles such as engines, ladders, rescues, squads, and ambulances. **Equipment** is defined to be the tools needed for the purpose, activity, or job at hand. Based on the threats to your community, you need to acquire the tools to do the job. Do you have a plan? What apparatus do you need? How long will it last and how often will it need to be replaced? Is it designed for your use? As with every other aspect of emergency service, know your community's needs and use your funding to obtain what you need to do the job. Do not buy vehicles solely based on tradition, what others may have, or what the vendors have to sell. Get what you need based on service demands.

apparatus
The "rolling stock," that is, vehicles such as engines, ladders, rescues, squads, and ambulances

equipment
The tools needed for the purpose, activity, or job at hand

You as the chief are responsible for the acquisition and maintenance of the necessary apparatus and equipment needed to do the job properly. These tasks involve budgeting, specifications, following purchasing policies, and supervising ongoing maintenance. Further, you are ultimately responsible for the daily regular equipment checks to ensure that vital equipment and apparatus are always ready.

Depending on the size of your organization, you may rarely use apparatus or equipment. It is not your job to function as the driver or engineer, and you will not be participating directly during the emergency. Thus, you need to involve those who will be using the apparatus and equipment in the acquisition process. As a result, better apparatus and equipment will be acquired, and it will be better maintained, because those who use it are involved in the process. This is a very simple concept, but many chiefs want to exert their authority in this area because it is what they know best and want what is important to them. Resist the temptation to do this. Your job, like so much else of what you do, will be to coordinate this process and let those with more expertise and ownership make the recommendations. Give them parameters with respect to funding and function and let them return to you the specifications that fall within your guidelines. Ultimately, you will make the decision based on the recommendations of your personnel, the budget funding available, and possibly some political considerations. Sales people, companies, or manufacturers with a tie to your community leaders may influence decisions made by the policy makers. Be mindful of this, though it doesn't always mean you can't obtain what you need.

Consider what emergencies are likely to occur in your community. Occasionally review the types that are most frequent and those that don't happen so often but do present a threat. Your preparation should reflect this. Sometimes organizations reflect more the national culture of the fire service than the needs of the local community. Because funding varies from community to community, your preparation will be affected by your community's ability and willingness to pay. If you are unable to acquire a certain piece of apparatus, for example, a ladder truck, it will affect the tactics you employ. Develop your program based on your local circumstances (see Figure 6.2).

◆ INCIDENT ANALYSIS

If your department is typical and has EMS responsibilities, they can constitute approximately 60 percent to 80 percent of your emergency responses. Consequently, this is a very important part of the job. Much of the preparation for this service is preestablished, as there are many laws, rules, and standards applicable to EMS. Depending on the level of service provided, whether basic first responder, EMT, intermediate, or paramedic, and whether you transport, much of what is required is dictated by state and local requirements. It is important to know what you need to do

FIGURE 6.2 ◆ Fire apparatus must be appropriate for the services provided.

to provide the highest possible level of care within your agency's level of licensure. Personnel must be prepared and then properly equipped. Decisions must be made as to what staffing and response model will be used. For example, do you have ambulances (sometimes known as rescue vehicles in some parts of the country, but referred to here as the typical "box"-style vehicle associated with EMS calls), or are personnel dispatched on fire engines or pumpers or other apparatus that is not an ambulance? If so, there are questions to consider regarding design and use of the apparatus. Do you operate in partnership with a private ambulance company? You also may need to license your apparatus in accordance with your state rules. Check these out to avoid potential problems.

Training for EMS is relatively straightforward. There are minimums required to obtain and maintain licensure. Personnel will need continuing education credits (CEUs), which will add to the already busy training schedule. This requirement may also add overtime costs, something to address at budget time.

Specific on-scene issues for EMS calls revolve around the chain of command and interaction with those not in the fire department, such as private EMS providers and law enforcement officials, particularly on the scene of injury accidents or in incidents involving mutual aid response, especially those that occur on your borders. Chain-of-command issues arise because of differences in department organization charts and established laws regarding medical care. In most cases the highest ranking medical authority has control of the patient. For example, a firefighter paramedic probably (most likely but make sure in your state) has authority over a company officer without a paramedic license with respect to patient care. Also, paramedics outside the fire

FIGURE 6.3 ◆ Interactions with other agencies during emergencies must produce good services.

department, either third-service or private company personnel, have patient control. All fire personnel need to understand the legalities so conflicts do not develop on the scene of an emergency. Law enforcement personnel have responsibilities for traffic management. Occasionally their views of situations are different from those of fire department personnel. As chief you need to make sure that everyone understands the protocol with respect to who is in charge ahead of time. You will need to meet with those outside your organization to make sure you are all in agreement. Disputes on the scene detract from patient care. Know the laws, rules, and policies so egos do not create problems on incidents (see Figure 6.3).

Although EMS is the most frequent service, there are many other types of incidents that departments must be prepared for. They present more challenges because they are potentially infrequent. Frequent incidents enhance skills, while those not practiced regularly will not be to the level needed or expected. You must make sure your organization and its people are properly prepared regardless of the expected frequency of occurrence.

◆ FIRE AND SPECIAL INCIDENT RESPONSE

Although most departments find that fewer than 5 percent of their responses are to actual fires, preparation is vital. Further, other types of calls even less frequent, such as those involving hazardous material, special rescue (swift water, high angle, ice, confined space, trench, collapse, etc.), wildland, and wildland–urban interface fires and

FIGURE 6.4 ◆ Special training is needed owing to the demands for more service and competence.

disasters (whether human-caused or natural) still require preparation. In some cases, the capabilities of the department may not be adequate to the task. Outside help must then be evaluated and considered (see Figure 6.4). There is no doubt that when an emergency happens, the fire department will be called. It may or may not have the resources to totally abate the emergency, but it will be expected to find the resources to handle the situation properly and safely. If the fire department does not have the resources needed, they will be found in other local agencies, through mutual aid, and/or as part of other governmental bodies, for example, county, state, and federal services. For example, a terrorist incident will require assistance from agencies at all levels of government. Incidents involving hazardous materials may also require assistance from the Environmental Protection Agency (EPA), state agencies, and possibly private specialists. Wildland fires may require firefighting or other assistance from states and/or federal forestry agencies. It is up to the chief to identify these resources and consider them in the overall plan to handle emergencies. Besides knowing the available resources, one of the biggest challenges is staying current with emerging trends, laws, and policies. Special rescue situations are dynamic, so be prepared.

◆ FIREFIGHTER PREPARATION

The fire service is a labor-intensive business. Most of what is done requires humans. Even though technology has helped, people are needed to do the job. As a result, firefighters need to be adequately prepared. The preparation starts with an assessment of the current situation. What are your training standards including minimum licensure and certifications requirements? Do you have regular training programs to maintain skills at a high level of **proficiency**, that being the competence of your personnel at their job? How do you evaluate that proficiency? What is the experience level of your personnel? Typically, incidents with a high frequency rate are handled most effectively. Those less likely to occur are more at risk for an unacceptable outcome. For example, communities with low fire volume may not have experienced personnel capable of making proper decisions. You do not have the luxury of waiting for enough incidents to allow personnel to gain the needed experience to properly discharge their duties and keep all personnel safe. If this is the case, it is up to you as the chief to develop a plan and establish direction to accommodate this potential deficiency.

proficiency
The competence of personnel at their job

When frequency is low, the only way to compensate is through training, education, and sound policies and procedures. Departments with low numbers of fires must train more, educate their personnel, practice, and establish appropriate and adequate policies. With respect to providing more training, it must be appropriate for the rank of the personnel. Firefighters get firefighter training and must drill repeatedly to maintain the necessary skills. Just like sports teams, musicians, or other skilled groups, they must practice enough to establish good instincts, which are so important during an emergency. The instincts can be developed only through practice and repetition. Of course, firefighters often do not like this because it is not exciting and can be perceived as boring. Do not let this prevent you from doing what is right.

Time can be one of the biggest challenges when establishing the necessary amount of training and repetition needed. The work schedule of career departments, run volume, and even labor agreements that have restrictions on training days and times (some contracts prohibit training on weekends and holidays and at certain times or under particular weather conditions) cut into the available time. To compensate, the training needs to be well organized. You or your staff can create quick training programs that can be done in 10 minutes or less. These can then be done at any time and with minimal supervision. Mostly these will be the basics—stretching hose line, placing ladders, wearing SCBA, ventilation, salvage, overhaul, forcible entry, apparatus placement, and so on. More advanced training or class room instruction will require more time. In these cases you will need to consider overtime options and/or taking units out of service. You and your staff will need to get creative so that every day is a training day.

FIGURE 6.5 ◆ Training is important for maintaining a high level of competence of "A" service.

Volunteer and on-call departments also have challenges with managing time. Again, it is important to focus on the basic needs of the organization, even though some of the training may be perceived as boring or repetitive. Good organization of your training will maximize the benefits. Usually, there are regular training nights. Make sure these are not wasted on nonessential issues. Also, the short, 5–10 minute drills may be of value after calls. You can often do this without adding a significant burden to your personnel. But remember that being good, whether or not you are paid, is a requirement and an expectation. Training and preparations are part of the job (see Figure 6.5).

◆ STRATEGY AND TACTICS

strategy
An overall plan

tactics
The means to accomplishing the goals of a plan

Officers must receive higher levels of training, exposing them to situations that they might face. They need to understand strategy and tactics, incident management, scene safety, special rescue requirements, and media relations. **Strategy** is the overall plan, and **tactics** are the means to accomplishing the goals of the plan.

Solid training is essential and it must be appropriate. Firefighters rely on the direction provided to them by their officers. Get everyone to the level of competence needed to excel at the job. Failure to do this places personnel, citizens, and property at risk.

The goal is to make everyone in your organization extremely competent at their jobs. This task is ongoing and never complete, so do not accept anything less than efforts to continually improve. Further, you have the responsibility to prepare personnel for future assignments. Do not create a situation in which individuals are promoted and then given on-the-job experience. They should have an established skill set prior to promotion, including qualifications and certifications, if available. Do not assume that good firefighters make good officers. Different preparation and skills are required. Your training program must work to enhance current performance and prepare people for advancement in the department with an emphasis on emergency scene responsibilities.

Even more challenging is the preparation needed for the less frequent incidents mentioned previously. The foremost challenge is keeping personnel prepared and motivated. With the necessity of finding time to train and prepare for ever-increasing responsibilities, it is easy to ignore or postpone training for the types of incidents that historically are not likely to occur.

Firefighters, like most people, tend to think that certain things are not going to happen to them (much like homeowners who don't think they will have a fire). Personnel often do not commit to preparing for an event if they perceive its likelihood to be very small. For example, hazardous material incidents are infrequent (at least the major ones). Proper preparation requires technical knowledge and practice with specialized equipment. Such preparation may not be high on the priority list of those who joined to fight fires. Firefighters may have a negative attitude toward training for something they perceive as not important or something they may not use. Further, the resource providers (politicians) may not be willing to commit the funding to purchase items necessary to respond to such infrequent events.

Often, only firefighter training is mandated. Although some states have requirements for fire officers, they may not be explicit in the details. OSHA rules require knowledge for the job being done. Training must be provided based on the job assignments. Officers need training in strategy and tactics. Command officers need training in incident management. (*Note:* The federal government, though not through law, influences this training through requirements attached to funding applications, grants, and the like. For example, the federal government strongly encourages the use of the National Incident Management System (NIMS). Failure to adopt and train to this requirement will, most likely, prevent your obtaining any federal grants). All line personnel need to understand all elements of emergency operations, with an obvious emphasis on their primary responsibilities. A plan to provide the necessary training must be developed and implemented. The plan should include the establishment of policies, delivery of training based on sound practice and established procedures, and an evaluation of the effectiveness of the training. Utilize the expertise in your organization and consider partnering with other fire departments that should be doing the same thing.

◆ PERSONAL DEVELOPMENT

Although the chief generally does not handle the day-to-day emergencies, there is an expectation that the chief will be an expert. In reality, you cannot be an expert in everything, but you do have the responsibility to surround yourself with good, solid people who can create a staff of experts. Personal development in this area is important not only for the routine but also the rare incidents. You will be relied on to be the expert or the ultimate source of expert resources, so you need to prepare. Attend classes, read articles, review incidents, and talk with others to keep your mind sharp, to develop networks, and to enhance your critical thinking skills. The types of incidents that will require your direct involvement are rare and often present unique challenges. You will need to think on your feet. Remember that there is usually a logical procession to all emergencies. Remember the basics—life safety, incident stabilization, property conservation, restoration. Your decisions should be based on the good of the community, the good of the department, and the good of your members.

Incident scene safety for responders must always be the major concern of the fire chief. A culture of safety needs to be established in your organization. This can require a significant shift in the thought process of your organization, depending on its history. Although emergency scenes are inherently dangerous, risks need to be minimized, and only in cases where human life is endangered should great risk be taken. You cannot be present at all emergencies, so you are reliant on the entire membership. Your role is to create a culture that mandates safety, delivers the necessary training, and provides the safety equipment to do the job. Policies and procedures must be in place and enforced all the time. Everyone must go home at the end of the shift and/or at the end of the incident.

◆ SUMMARY

This chapter discussed some of the challenges of emergency response you as the chief must face. Remain vigilant and accept these challenges as part your job. Personnel and policy makers must be motivated to do their part. If you and they are not prepared and an incident does occur, you will be the one criticized and asked to provide answers. The core mission remains response to emergencies and chiefs are expected to be perfect or near perfect every time.

ACTIVITIES

1. Survey the apparatus used in your community. Discuss their applicability to the hazards faced in your community.
2. Identify the most common types of emergencies likely to occur in your community and the less frequent ones. Compare the preparations needed for these various emergencies with the abilities of the department to address them.
3. Evaluate your EMS system in your community. Identify the model currently being used. Develop a plan to enhance the quality of EMS delivery.
4. Identify the NFPA standards applicable to your department. Evaluate your level of compliance.
5. Research the applicable laws in your state relative to emergency response, including OSHA standards. Determine your level of compliance.
6. Identify your current methods of providing training for major incidents. Note

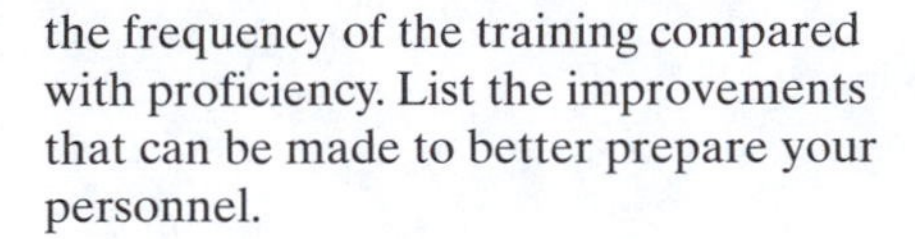

the frequency of the training compared with proficiency. List the improvements that can be made to better prepare your personnel.

7. Review the requirements of the National Incident Management System (NIMS), and evaluate your department's level of compliance. Look more broadly and evaluate your municipality's level of compliance. Develop a plan to bring your department, and perhaps your municipality, into full compliance.

CASE STUDY

PROBLEM WITH MUTUAL AID

Your department is dispatched to an injury accident on the border with a neighboring community. Your closest units are already operating on other incidents. Your on-duty shift commander asks that the next available unit be dispatched and also requests a dispatch from the neighboring community, since it has a station considerably closer. Your shift commander arrives on the scene and is immediately asked by the officer from the other department why its unit was dispatched to your community for a "simple" accident. This confrontation occurs in front of the patients, and the other department's members are disrespectful to your personnel. Ultimately, one of your units arrives and takes over care of the patients, and the mutual aid company is released. This incident is relayed to you the next morning when you arrive at work.

What will you do?

CHAPTER 7

Routine Issues

KEY TERMS

delegate, p. 88
distraction, p. 79
document, p. 85
interruptions, p. 78
meetings, p. 82
open-door policy, p. 87
rumors, p. 88

Too often when considering a job and its duties, we overlook the routine, normal, mundane, and sometimes boring day-to-day requirements. We focus on the unusual, exciting, or major projects and don't always commit the energy necessary to do the things that have to get done on a regular basis. The fire and emergency service is very unpredictable, which is what makes the job attractive to so many people. So much of what happens is exciting and can change dramatically from day to day, including both emergency and nonemergency events and incidents. However, much of the work done by the fire chief is predictable and vital to the overall mission of the organization. Think of the components of a chief's typical day—meetings, correspondence (both incoming and outgoing), payroll, invoices, purchasing, oversight, supervision, short-term planning, and so many other minor things that they don't always register. The competent executive learns how to handle these issues while being prepared for the inevitable unexpected events.

interruptions
Breaks in activities that temporarily halt work

You as chief must learn to deal with constant **interruptions**—personnel, the telephone, citizens, your boss, e-mail, and the like. These interruptions are breaks in activities that temporarily halt your work. Regardless whether the interruptions are good or bad, you can expect them to happen. A good skill to develop is the ability to work in short bursts to accomplish tasks and be able to move on.

Seldom will you have a large block of time in which to accomplish projects. You need to develop a system that allows you to change gears quickly and still complete assignments. Often, you will have to multitask, that is, keep a lot of balls in the air and change them as needed. Many activities rarely last long, yet it remains important to give them the proper attention, paying attention to detail and avoiding mistakes. Much activity revolves around paperwork critical to the organization. Other activities involve personnel, and as with so many issues, relationships become very important in dealing with day-to-day responsibilities.

◆ DISTRACTIONS

One of the most important things to learn in order to be effective is to control distractions. This is vital, because distractions are a certainty. A **distraction** is different from an interruption in that the former is defined as anything that interferes with your concentration or takes attention away from something else. Most of what you do requires your undivided attention, so distractions can cause serious problems. The trick is to minimize their impact on your ability to do your job while allowing access to those around you. A great support staff and/or secretary can help in this regard. They know when to interrupt you and when to protect you from calls that are not as important as the project you are working on. Your secretary or whoever answers the phone should know the people who should always be able to get through (your boss, your children, your spouse, though not necessarily in that order!). You also need to tell the person answering the phone when you are expecting a call or under what special circumstances to pass on calls. If the call is someone you must speak with, do all you can to accept the call. With the prevalence of voicemail, a missed telephone call can lead to numerous calls back and forth between machines until you actually contact the person. Controlling this game of telephone tag by having your staff know when to alert you is important but does not eliminate the need for you to personally control the interruptions (see Figure 7.1). Again, remember that controlling is a fine line. Do not create the perception that you are in an ivory tower. If people can never reach you, maybe you need to reevaluate your priorities. You need to maintain relationships or you will find yourself with even more distractions.

distraction
Anything that interferes with concentration or takes attention away from something else

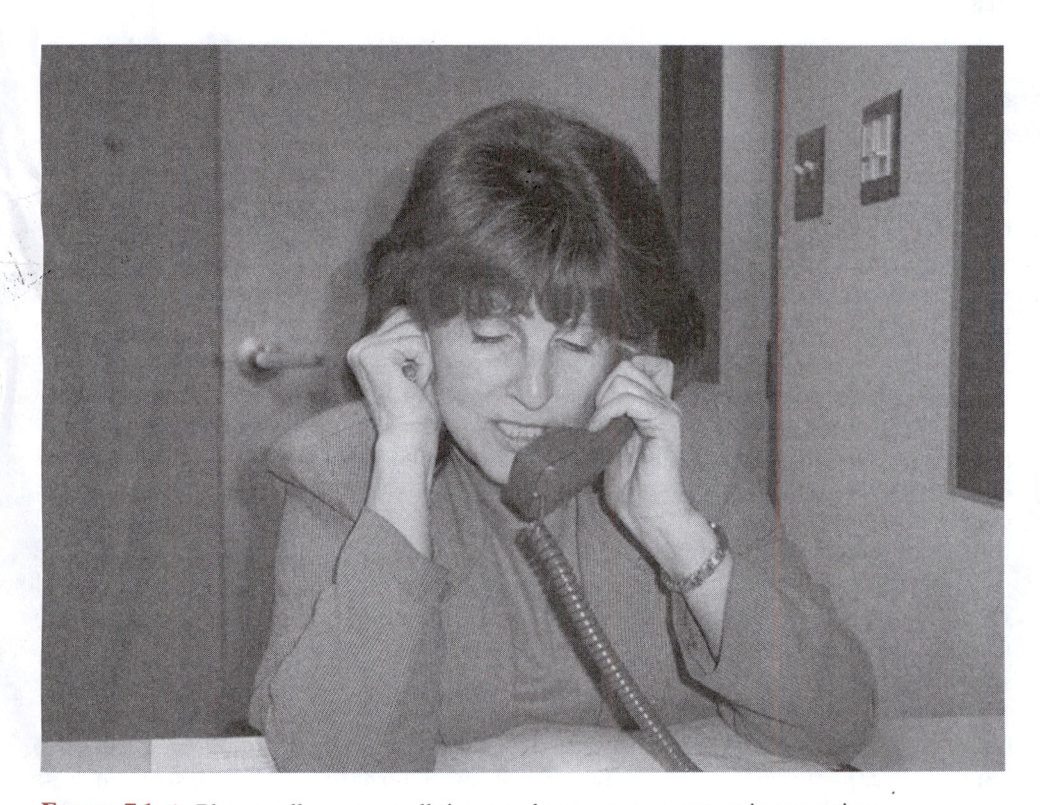

FIGURE 7.1 ◆ Phone calls come at all times and can create constant interruptions.

This may sound contrary to the preceding advice regarding the control of telephone calls, but you should make every effort to return calls as soon as possible. Except in rare cases when there is a reason to delay getting back to someone, such as the need to do research, a rapid reply makes good business sense. With today's technology, a call can be returned at any time. Cell phones make it easier. Plan your return calls around downtime—in an airport, driving (if you have hands-free capabilities and it is safe to do so), or when waiting for meetings or appointments. You will find that making a quick return call will not only help with relationships but will often solve a problem, prevent more work, or avoid other time-wasters.

◆ ORGANIZING FOR PRODUCTIVITY

Next, have a plan. How is your day organized? Do you have a system for doing the normal functions? Determine whether the beginning or end of the day is best for you. Some chiefs come in early to do things before others arrive and the phone calls begin. Others find the end of the day more conducive to their work style. Whatever system works best for you, set time aside at either the beginning or end of the day. The purpose of this time is to sort and read mail, review and sign documents, draft reports and correspondence, and other tasks that are best done alone and without distractions. This can also be the time to do the work that requires more concentration, as there will be fewer interruptions or distractions. (The author has found that there is no perfect time consistent with this goal. Someone or something will always find a way to disrupt your plan!)

◆ TECHNOLOGICAL ADVANCES

One of the biggest changes in time management in recent years has been the advent of e-mail. Although it is a great tool for improving communications, it can also create some challenges. You can easily get 100 or more e-mails a day. If each one takes you just 30 seconds on average to open and scan (realizing some must be read, while others should be deleted immediately), you can count on a minimum of 1 hour added to your work day. Avoid getting on group e-mail lists—ones where your friends forward everything they get plus material they think is really funny. You may get an occasional good e-mail that you really appreciate, but the time saved will make it worthwhile to get off the list. Another potential pitfall with e-mail is to get too attached to your desk and forget the other important means of communication, such as face to face and the encounters created when you practice MWBA (see Figure 7.2). You can get tethered to the computer if you are not conscious of what is happening. You know you have gone too far if you send an e-mail instead of walking to the office right next to yours! Finally, work with your computer experts or information technology personnel (IT) to reduce spam, the unwanted and unsolicited e-mail. They are experts on the many software programs available.

Cell phones are another modern tool for communication. As mentioned previously, use a cell phone during downtime or while driving, if you are allowed and have hands-free capabilities, to make or return calls. An alternative is to have someone else drive so you can take care of business. Between your cell phone and your office

FIGURE 7.2 ◆ Computers can save time but must not replace important interactions with people.

phone, you should always be able to return calls. Calls that are not returned can create relationship problems. Always return calls as soon as you can.

Technology has also given everyone the capabilities to send and receive e-mails wherever they are. Through devices such as the BlackBerry® you now have the capability of instant communications with e-mails. As with any technology that promises to make life easier, there are both positive and negative aspects to consider. First, if you are typical, you receive many e-mails. The ability to answer and/or delete them wherever you are is a time-saver. If you are gone from your office for any length of time, you do not want to return to a large number of notes on your computer. You also have the ability to respond to time-sensitive issues. On the downside of the technology, other people will now expect instant responses from you. You can also rely too heavily on the device. As with all "time-savers," search for balance. Do not neglect the other important aspects of your job, that is, personal relationships and the value of direct interactions. Also, do not let the device distract you when you should be paying attention to something live, such as a meeting.

There are a couple of caveats to remember as technology continues to take hold and become more significant to the job. Always remember that e-mails can be subject to discovery in legal proceedings and also may be requested using the Freedom of Information Act (FOIA). Never put into an e-mail anything you would not be comfortable seeing on the front page of the local newspaper. The other technological advance that you must be aware of is the digital camera, both still and video. Cameras have proliferated, and it is best to assume that your photo and actions can be recorded and can appear on a video or photo anytime you are outside your own home. Be conscious of your actions at all times. If you are conducting yourself as the professional you are, this should not be an issue.

◆ TIME MANAGEMENT

Develop a sense of timing, that is, the ability to know when to act. It may also mean knowing when *not* to act. There are many "crises" that cross your desk. Often they resolve themselves if left alone. Real issues require research and investigation and do not require immediate action. Take your time. Do what is right. Few things are truly life threatening. Talk to others in your network to help you figure out what needs to be done.

Do your best to prevent people from controlling your time. You can set the schedule with most people with whom you interact, but realize that there are some who will have the authority to control your schedule to some extent. Set aside time for events beyond your control so that if they happen you are ready; if they don't, then you have extra time. Either way, you are prepared. When you plan, anticipate that some things will be out of your control. Still, you can pick your spots, depending on the job at hand. Meet with people whenever possible, even if it messes up your schedule. Each time you do this, you create goodwill, which will allow you some flexibility when you really need it. Remember, too, that relationships are important here. When you can have open and candid conversations and trust exists, you can request some leeway when others in authority want your time. Some chiefs remain at the mercy of their boss and get very frustrated at the demands and last-minute requests. These types of situations can add stress to the job and make time management more challenging. When you continue to build relationships with the boss and others, the working environment improves, which will allow you more control, even with those in positions of authority. In these cases it is vital to know what is important to the boss and others with more influence over the fire department.

When your time is under the control of others, always be prepared. Take work with you. When a meeting is delayed, you can catch up on some of your routine tasks. You can take advantage by reading some of the periodicals of your profession, reading letters, writing hand-written notes, proofreading documents, or answering e-mail on a BlackBerry® or similar device. There is always more work than the time to do it. Always have something to do in case you get the opportunity (see Figure 7.3).

◆ MEETINGS

meetings
Occasions where people gather to discuss something

Meetings are a routine part of every executive's life. They are occasions where people gather to discuss something. You cannot avoid them. They will occur and you will be expected to participate. Schedule the ones in your control around your own schedule. Leave time for others that may be out of your realm. If you are in charge, consider a written agenda (see Figure 7.4). Let everyone know what is to be discussed. If other issues arise, start a list for a future meeting and stick to the topic.

FIGURE 7.3 ◆ Everyone has the same amount of time to do the job.

BOARD OF CHIEFS

MEETING

Monday, March 19
1500 hrs
Conference Room/Station 5

Agenda

I. Call to Order
II. Approval of Agenda
III. Approval of Meeting Minutes
 a. No February meeting minutes posted
IV. Unfinished Business
 a. Payroll
 b. Smoking
 c. QI reports
V. New Business
 a. Uniforms
 b. BLS patients
 c. Thermal imaging camera
VI. Staff Reports
VII. District Chief Reports
VIII. Adjournment

FIGURE 7.4 ◆ Use a meeting agenda tailored to your needs to let everyone know what is to be discussed and to keep your meeting focused.

Keeping minutes of the meeting may keep you from "repeating" the meeting later on with others or those with short memories. Minutes also help with communication to individuals who were unable to attend and remove the mystique of the meeting, which helps control the rumor mill. Minutes can also remind people of issues that have been discussed and resolved.

When you do not control a meeting, you can be stuck. This comes with the job. No one said it would be fun all the time! If it is a general meeting, take some paperwork or reading as long as it is not disruptive. You may be asked to attend city council meetings regularly, at which much non-fire-related business is usually discussed. If you take some reading, it might help clear some things off your desk. An alternative is to send others from your staff to participate (if allowed). Your department is represented, the information is obtained, and you can use the opportunity to develop your staff and expose them to other facets of the job.

You can view meetings in a number of ways. They may be boring, unproductive, and time-wasters, but they are an important way to share information. They also provide a chance to interact and build relationships with others vital to the mission of your organization as well as demonstrate your ability to be part of a team. Productive meetings can be short or lengthy. Sometimes the short meetings are best. Meetings can be about regular issues, such as weekly staff meetings, or for a specific purpose (discussing a project). Invite only those who need to be there. Some people are meeting killers. They talk too much, get off subject, interrupt others, keep others from participating, or create a less-than-professional atmosphere. In most cases you can control the members of your department, but in other circumstances attendees are not under your command and control. See Figure 7.5.

FIGURE 7.5 ◆ Meetings are a part of everyday life for a fire chief.

WRITING

As the fire chief you create a lot of documents, You need good writing skills, because you may be judged on the documents you produce. Consider that a **document** is a formal piece of writing that provides information or that acts as a record of events. Regardless of the method of writing, be it a letter, memo or e-mail, all are important. It is hoped that you remember your writing lessons from school. When you produce a document, make sure it is written properly with correct spelling and grammar. Today's computers have spell and grammar check, which reduce but do not eliminate errors. The software can detect whether a word is spelled correctly but not that it is the correct word. You also should have someone such as a secretary or staff member who will read your documents and offer help. You should also be able to distinguish between formal documents that need to be perfect or near perfect and those that can be simple communications. Spend time on the important ones. You can start with an outline, brainstorm your ideas, and then organize them as best you can. Read and reread to make sure you are saying what you intend.

document
A formal piece of writing that provides information or that acts as a record of events

The more important the document, the more time you must spend to make it as precise as you can. Simple, succinct documents are preferable because they minimize the chance of misinterpretation. Most people do not have the talent to waver from basic written communications, so keep your writing style simple and straightforward so that it correctly states what you want to convey. Poor writing can mislead people or even produce hurt feelings. Be careful, reread, and if you can, wait a day before sending off a document so you read it with fresh eyes the next day.

Some people are capable of creating their own documents (either by typing or word processing). If you are capable and have the time, good for you. Others rely on a secretary to produce the documents. Secretaries are by training probably faster and more accurate, and they supply a second set of eyes for reviewing the document. A secretary can work from a written draft or dictation. The choice is yours. If you are dictating, there are a few suggestions.

- Start with an outline.
- Speak slowly and clearly.
- Spell proper names and uncommon words.
- Insert punctuation.

If you intend to review and edit the document, have it double-spaced to allow for comments and changes. Allow your secretary to edit, suggest changes, clean up language and punctuation, and offer improvements. If the document is really important, have someone else read it before sending it out. Be especially careful if documents involve discipline or if you are reacting emotionally. Documents related to disciplinary issues must be perfect or as close to perfect as possible. Someone in the department should review them, as well as one of your experts, either the human resource director or labor attorney. In cases where you have the urge to respond instantly, let the document sit overnight, then reread it and see if you still want to send it. Often, you will change the document or choose not to deliver it.

POLICIES AND PROCEDURES

Writing policies and procedures is not the most exciting part of a fire chief's job, yet you must do them. Often, you can delegate this assignment to another person or a committee, but there will still be times when you need to do it. There needs to be an established format that provides a template (see Figure 7.6). This template will include the essential elements of all procedures: the topic, who it affects, who is responsible, the desired behavior, the effective date, and any consequences for failure to abide by the policy. They need to be simple and to the point. Ones with special importance and potential legal ramifications should be reviewed by either an attorney or human resource expert or both.

SAMPLE PROCEDURE

FIRE DEPARTMENT

SAFETY PROCEDURE

PRIORITY: 1

USE OF SCBA AND PROTECTIVE EQUIPMENT

NO: 910.0

EFFECTIVE: 06/01/03 PAGE: 1 OF 1

(Rescinds version: 11/15/95)

FIRE CHIEF APPROVAL:

PURPOSE:

The purpose of this procedure is to identify the proper use of self-contained breathing apparatus.

PROCEDURE

I. Protective Equipment Usage

A. Protective clothing and self-contained breathing apparatus (SCBA) are provided for the well being and safety of each firefighter. The responsibility of using it properly rests with the individual and his/her immediate supervisor.
B. Self-contained breathing apparatus shall be worn by all fire department personnel working in areas where:

1. The atmosphere is hazardous, or
2. The atmosphere is suspected of being hazardous, or
3. The atmosphere may rapidly become hazardous.

C. In addition to the above, all personnel working below ground level or inside any confined space shall wear SCBA unless the safety of the atmosphere can be established by testing and continuous monitoring. These situations shall include contaminated or oxygen-deficient atmospheres caused by a fire or other emergency.
D. Personnel reporting for assignment on the fire ground or other potentially dangerous emergencies shall assume the need for SCBA and have in their possession an SCBA with face piece. They shall be outfitted in full turn-out equipment including boots, bunker pants, coat, hood, gloves, and helmet. When responding on apparatus, members shall be in proper apparel prior to departure. No delays in operation should occur because of failure to wear proper protective clothing.
A. This policy shall apply to all fires, all phases of fire fighting, and any hazardous material incident where toxic fumes may be present. The use of SCBA shall continue until the incident commander determines otherwise.

FIGURE 7.6

MISCELLANEOUS

Common courtesy is always a good practice. Remember to say thank you every chance you get. Make it a habit. Write notes when appropriate. This can be to members, citizens, the business community, or anyone who does something to warrant an acknowledgment from you. Sometimes a handwritten note from you is most appropriate. Though these may seem insignificant to you, they are likely to be remembered for a long time by the receiver. Other times a formal response is the best approach. Others may be handled with a simple thank you in passing. Regardless of how you say thank you, be specific. A generic "thanks" is not nearly as meaningful as one that mentions the specific reason for the appreciation. For example, instead of saying "thank you" tell the individual that you really appreciate the quality of the training program and the details that went into developing the content. Finally, if the note is to one of your members, place a copy in the file. It helps to remember the good things, too!

You will be asked to participate in many functions and rituals. Do not view these as a waste of your time. They are important for political and relationship reasons. As a result, anticipate that there will be a certain number of these and accommodate your schedule accordingly. These may include the following:

- Hiring and promotion ceremonies
- Awards
- Community- and business-based events
- School functions
- Other governmental department functions
- Ribbon cuttings, dedication ceremonies, and grand openings
- Retirement parties

Remember, these are all part of the job and must be accounted for in your planning and time management. They are important for many reasons. They create opportunities to build relationships with many different people in a very relaxed, informal setting. They further connect you to your community and let others know their events are important to you and the fire department. Your attendance may even encourage others to come to your events. Put on your best outward appearance and enjoy the experience. These responsibilities come with the bugles.

◆ OPEN-DOOR POLICY

Many executives have an **open-door policy**, which refers not to the physical opening of the door but to free and unrestricted access. This is a good policy, but it can create problems if not controlled. The benefits are that your members will feel comfortable talking to you about department issues. You can learn of things before they become a problem, and you can hear ideas about improving the operation. You can also hear about someone's summer vacation! On occasion the open door is an invitation to talk about anything. Although it is good to be personal and build relationships, know when to hold the line and stop the discussion. One way to get people out of your office is to stand and move to the door. With others you can be more direct and ask them to leave. Know your personnel so you know which tack to take. You don't want to offend anyone unintentionally.

open-door policy
Free and unrestricted access

Another possible problem with the open door is that it can undermine your staff. If people feel they can go around their immediate supervisor to get a sympathetic ear, you may lose credibility with your staff. You can deal with this problem by asking whether they reported the issue to their boss. If they did, you can hear them out then ask your staff for their perspective. On the positive side, this gives you a chance to clarify your position to your staff on a real situation.

◆ DELEGATION

delegate
To give someone the authority to act on your behalf

Depending on the size of your organization you will need to **delegate** many things, that is, give someone the authority to act on your behalf. It seems so simple and in most cases it is, but you must know your personnel. Everyone brings a skill set to work, and everyone has strengths and weaknesses. Assign the work to the person who will best do the job (bearing in mind the tendency to overwork your high achievers). When you delegate, give specific directions if the project calls for particular content, and let the individual know where he or she is allowed to provide individual insight and opinion.

Micromanagement is not appropriate when delegating. Make the assignment and allow the individual the freedom to choose the best way to accomplish the objective. However do not confuse micromanaging with your ultimate responsibility to know what is happening in your organization. Let everyone who may be on the receiving end of your delegation know that when you ask questions and learn details about the project you are not micromanaging but only fulfilling your own responsibility. Looking over someone's shoulder after you have provided specific directions can be overcontrolling. Know the difference and communicate your expectations. Also provide a due date or time frame. If the project is important and time sensitive, say so.

Remember to delegate the issue to the level of the organization where it can best be addressed. When delegating, it is important to professionally approach the individuals being asked to take on the task. If the work, task, or project is not done to your standards, you need to provide additional direction. Unless there was a willful disregard of your direction, you need to communicate your desires again and make sure the person understands. In doing so, maintain your composure, display the proper demeanor, and plan to follow up to ensure success. If you have done all this, you can hold those involved accountable for their part.

◆ RUMORS

rumors
Idle speculation or unverified reports

Rumor control seems to be a major function of the fire chief. In the absence of information, people will make it up. Some do it for fun, some do it because they like to talk, others have partial information and want to impress others with what they know, and some are actually malicious in their actions. **Rumors** are unverified reports or idle speculation; they may or may not be true but can be very destructive to the organization. You need to take action to try to control any misinformation that appears in your organization. You do this through communication, using every means at your disposal, including memos, newsletters, meetings, minutes from the meetings, phone calls, e-mails, and anything else that will get the correct information out. Regardless of how many means you use, some people just won't get it. Keep doing what you can.

Another way to address rumor control is to seek out the source and confront the individual. Use the opportunity to educate and inform him or her about the dangers to the organization regarding false information. If the rumor is malicious, consider disciplinary action. Whatever you do, send a message that false rumors will not be tolerated. Create a culture that lets the organization know that you are more than willing to give out information regarding the operation of the department.

◆ OTHER CONSIDERATIONS

How do you dress? First impressions are critical. People often begin to size you up long before you start talking. Some chiefs believe they need to wear a uniform every day. Others dress in business attire, leaving the formal wear for special occasions. Whichever you choose, be aware of the image your dress projects. What is the expectation of your boss or the community? Some feel that not regularly dressing in uniform produces a greater impact on those occasions when the more traditional garb of a fire chief is worn. Don't do what has always been done. Create your own image.

As chief you will be counted on to make decisions. Other than on the scene of emergencies, you almost always will have the time to consider the facts. Still, sometimes you will need to make quick (not necessarily instant) decisions. Develop a method to do so. One suggestion is to ask yourself a few questions: What is good for the community? What is good for the department? What is good for the individuals involved? There may be other considerations as well. Just be prepared.

It is a good idea to always have your résumé and a short bio ready and current. This is not for your next job (though you may want it at times!) but to be ready should there be a request for it. For example, you may be asked to make a presentation, and the host will want to introduce you, or you may be involved in a court case or arbitration. Your qualifications are important and a list of them needs to be readily available. Update the information regularly, and have others review it from time to time to make sure it includes the important items. You may also want to keep a file of your accomplishments and training and use it to create your résumé or bio.

Because of your position as fire chief your words carry a lot of weight. Everyone has opinions on different matters, but as chief, you cannot always offer your opinion. Also realize that verbalizing a negative opinion in front of the wrong people helps no one. You do not need to comment on everything. Know when to bite your tongue, and stay out of issues where sides may be developing. Be mindful of the power of your words. You are expected to have a positive attitude and be loyal to your organization and community.

You have many advisers offering to help you—human resource directors, attorneys, finance people, IT staff, and others. They are valuable resources, but remember that the fire department's "buck" stops with you. The people mentioned can provide information for your consideration, but you make the ultimate decision. There is a reason you are the chief. Think, consider, and act. Do what is right based on the information you received from others and your knowledge and instincts. Do not let others make decisions for you.

How do you cancel ineffective programs? This is a difficult exercise, because some have become institutionalized, so they continue without regard to logic or effectiveness. Some are out of your control. Others are relatively minor to you but important to others. In these cases, it might not be worth the fight to change things. As with so many other things discussed, remember that you have the final say. Rely on your talents, abilities, and instincts to know when to act and when not to. Rely on your communication skills and relationships. Do not simply issue an order, but plan the process to change.

◆ SUMMARY

Day-to-day routine activities require organizational skills, time management, tact, timing, and patience. Do not underestimate the importance of your ability to handle the basics of the job, as they may be the criteria by which your success is evaluated. Your day can get extremely busy, and some days you wonder where the time went. You are rushed from job to job, and everyone wants some of your time. You may not know how you will get your work completed. Do not get frazzled. You must keep your composure and keep moving on. People watch how you respond to pressure. Fire chiefs are expected to remain "cool under fire" not only on the fire ground but in the day-to-day performance of their job. Finally, maintain your enthusiasm. It is contagious and good for the organization as well as being a characteristic of most successful people.

ACTIVITIES

1. Develop an agenda for a routine staff meeting in your department. Include the topics, participants, and parties responsible for each topic.
2. Write down how your typical day is organized. Compare what you actually do with what you should do. Organize your day based upon your regular work. Arrange the order to be most efficient and complete the easy stuff. Set aside quiet time to get most of your critical work completed and time to return calls and written correspondence.
3. Identify five sources of management information that do not originate in the fire service (no fire service journals). Evaluate your choices to determine those that would be most beneficial.
4. Draft three documents applicable to your department: a simple memo, a formal letter to a citizen, and a disciplinary letter based on an infraction of your rules and regulations.
5. Deliver (or write) an impromptu speech to be given to the local social club about the duties and responsibilities of the fire department.

◆ CASE STUDY

SCHEDULING ISSUE: FAILURE TO COVER A TRADE

One of the basic premises of providing service is that you will have your stations staffed in accordance with your department policy. Your department allows unlimited shift trades within the preestablished guidelines agreed to by labor and management. On this day, the scheduled employee does not show up for work at the start of the shift. When contacted, the firefighter states that the shift was traded to another firefighter. When that firefighter is contacted, he denies knowing about the trade or agreeing to it.

Your shift commander has requested your advice on how you would like this situation handled.

Marketing, Media Relations, and Customer Service

8 CHAPTER

KEY TERMS

customer service, p. 100
effective, p. 97
efficient, p. 97
marketing, p. 92
media, p. 98
paradigm, p. 93

As the chapter title suggests, there is a relationship among marketing, media relations, and customer service. Individually, these areas contribute to the perception that people have of the department and assist in disseminating information and safety messages. These are not always thought of as topics of importance to fire departments (as indicated by the lack of resources committed to the job, including funding and training of personnel), but they are. In the past, perhaps, fire departments didn't need to be concerned with these issues, but now, citizens are more demanding and have higher expectations in return for the taxes they pay. Unfortunately, they have little time to investigate the services provided by their fire department or research, learn about, or understand the emergency service in general and the costs associated with them. They need to be informed so they can make appropriate decisions regarding taxes and government spending. You cannot expect the members of the community to get your message or understand what you do unless you continually provide them information about your service. In the end, they will support the services that they perceive to be important, efficient, and effective.

Marketing, media relations, and customer service all involve influencing the public's opinion regarding the department. The emergency service that you provide is a "product" that people pay for. If you want the "customers" to keep buying, you need to provide value, develop loyalty, make sure that they know what they are getting for their money, and create satisfaction with the service they may have received.

◆ WHY?

There are many examples of disconnects between the modern fire service and the people who are served. If you don't think so, ask some people you know—your family, your neighbors, or others you interact with. Ask them what they know about the fire department and the services you provide. Ask them if they know the qualifications of the firefighters. Ask if they know what it takes to be a paramedic. If they are unaware, then it is an indication that the general population is as well. Besides the questions you may ask others, pay attention to the questions you get, whether during informal conversations or when people are visiting the fire station. They may ask, Where is the fire pole? Where do the firefighters sleep? Where is the dog? What do you do when not on a fire? and others related to their understanding of the fire service. Their perceptions can be based on Hollywood stereotypes or the books they read as a child or someone they know who served a long time ago.

The fire department is a well-kept secret. There are many reasons for this, with the key one being a lack of self-promotion. Although it is not bad to be humble, there are consequences of inadequate advertising. In contrast, some departments are well known and have a great reputation. Think of those you know that are considered to be doing a good job, and ask yourself how you came to that conclusion. Was it through direct observation (did you actually see them perform?) or through reputation? If it was the reputation, how was it gained? Answers to these questions may help you understand the need to promote your organization.

Clearly, people want value for their tax dollars. They also want services that contribute to the community's quality of life, including its health, safety, and welfare. But in most cases, the public is uninformed or misinformed about what is available in their community and what is needed to provide the service they desire. For example, few understand the staffing requirements for a career department that is operational 24 hours a day, 7 days a week. Considering leave time, training needs, and other requirements and demands of the job, the department may require four or five firefighters to staff one position around the clock. Adding in the cost of a firefighter results in an expense that few realize is as large as it is. Consequently, the public, those buying your product, don't understand why you don't have enough money. The bottom line is that a failure to market will result in inadequate resources to do the job. Even with good marketing, you may still not get all you need, but you will certainly be better off.

◆ IMAGE

marketing The selling of a product or service

To engage in marketing and promotion, you need a product or service and an image. **Marketing** is quite simply the selling of a product or service. You know the services you provide, but what is the image you project? Often, this image is dictated by national or international events. Certainly, the terrorist act on September 11, 2001, and the subsequent response of the Fire Department of the City of New York (FDNY) established a positive image for firefighters. All fire departments benefited from the tremendous response and sacrifice made by FDNY. Citizens, in general, associate all firefighters with their portrayal in the media in the response to the terrorist attack. Likewise, acts of indiscretion can affect the entire fire service. Remember that reputations can be made individually or collectively—many people cannot distinguish one fire department from another.

What is your local image? How is your fire department viewed? Does it match your expectations, that is, how you want to be viewed? In many cases the image is one of a service that responds quickly, is well trained (especially in the area of EMS), physically fit, clean cut, and nice to people. This is not a bad image to have. Many a product or service would be very happy to enjoy that status. Marketing, public relations, customer service, and related programs are made much easier with such a great image.

As mentioned earlier, those who fund fire departments and call for service do not have time to learn what it takes to prepare for and respond to a variety of 911 calls. They just expect the fire department to be there when they call, act professional, and be really good at what they do, regardless of the reason for the call. It is easy to be overlooked in today's busy world. If you do not make the effort to educate and inform, people will call on their own experiences and interactions with the fire department, regardless of when those took place (yesterday or 20 years ago). That experience will be your image. If you haven't changed, then everything will be fine. However, if your department is much different than it used to be, you will need to let people know that. Further, partly owing to the decrease in the number of serious fires in most communities, fire departments are not always thought of in discussions of public safety. Today, many people associate public safety with the police department because of their exposure to "public safety" and their view of the world.

There are many good causes and many uses for tax dollars. There is no doubt that the fire and emergency service provides a valuable and necessary service for the health, safety, and welfare of the community, but other services and causes are often better organized and better at influencing public opinion and the policy makers of the community. Further, the public may still have the same **paradigm**, or typical example, of the fire department that has existed for years—that of a suppression unit, waiting for the next fire so they can rush off and save lives. If the number of serious fires is declining, then requests for adequate funding will face greater challenges. People are not likely to change their opinion without an assist from those in the fire service.

paradigm
Typical example

◆ MARKETING AND ADVERTISING

Think of those who market and advertise. Today, everyone does, even those who have not traditionally done so in the past. Hospitals, doctors, dentists, lawyers, school districts, utilities, and others have realized that they need to reach out to their customers. The outreach is designed to educate regarding the services offered, change opinions, and ultimately solicit support (i.e., money).

To start marketing, you need a plan. First, ask yourself a few simple questions: What do you plan to market? Who is your audience? How are you going to do it? Who is going to do it and will there be any partners? How will you fund it? Your plan needs to include the answers to these questions. It needs to be a systems approach, utilizing all available tools to accomplish your goals. Remember, marketing is not a one-time occurrence. If it were, companies like McDonald's, Ford, and Pepsico would not be spending the money and time they do on the many media outlets they use!

Huge marketing budgets are not the norm in fire departments, so a plan must be developed to utilize existing resources and opportunities. Your best resource is your people. They portray the image of your organization and have the potential to promote your services to those who need them. The most important thing they can do to help is to perform at the highest possible level every time they are asked to interact with the public, whether at an emergency response, a nonemergency response, or a scheduled event. If you can add value to those services, so much the better. People will remember, and if the event is significant to them, they will tell many other people. They will do this regardless of whether their experience was positive or negative. Hence, it is necessary to perform extremely well each time and to do whatever can be done to enhance the "customer service." Remember, the consumers' perspective is quite simple: they want fast response, nice people, and professionals who know their business.

◆ EVERYONE'S JOB

What are you doing to prepare your personnel to perform these functions? Too often fire departments do not commit time or training to marketing, media relations, and customer service. These have usually been learned from a predecessor, so that if the lesson was positive, the department does well; if not, then bad habits get passed on. Normally, marketing, media relations, and customer service are just expected to take care of themselves, and the product (or in this case, the service) is enough to convince the public that all is well. In many organizations these functions are not perceived as part of the job. Further, if they are, they must be the chief's or upper management's responsibilities. Obviously, this is not true. To be successful, everyone must participate, but it is the chief's role to get everyone involved and give them the tools to do so. This

FIGURE 8.1 ◆ There are many ways for the entire department to contribute to the marketing and image of the department.

means establishing policies and training personnel. It also involves the "selling" of the concept of marketing and customer service to all members of the department. Everyone needs to understand the importance and value of their individual performance to the overall effort (see Figure 8.1). They also need basic training on things that can be done to add value to everyday responses and routine assignments. Often this is accomplished by on-the-job training that is passed from member to member. If the "teacher" is good, then the right message will be relayed. Perhaps a better approach would be to include these elements in your training program, from recruit school to officer training. There also needs to be continuing education to make sure that the message stays current and focused. Finally, everyone needs to be empowered to enhance the level of service and deliver to the public the highest possible quality to ensure their loyalty to the fire department.

It is obvious that everyone needs to be considered a resource in the overall marketing of the fire department, but the role of the fire chief is paramount in establishing the mindset and plan. Studies have shown that the chief executive of a company may be responsible for up to 50 percent of a company's image. That CEO is you, the fire chief. Your image and professionalism go a long way toward establishing the credibility and capability of your department. Be good at what you do, be the expert, and be on your "game" every day. You are being watched and evaluated, and the expectation is that you will perform. Perceptions are developed based on short, single interactions, so you rarely have a second chance to change an opinion.

WHAT AND TO WHOM?

Of course in order to sell, you need to know what you are selling. Start by listing the things that you do for the community. Some are obvious—fire suppression, prevention, EMS. Others may not be so well known—disaster services, terrorism response, special rescues, outreach programs. In general, fire departments save lives, protect property, and provide whatever service they may be asked to do. Know everything you do and do what you can to make sure everyone else knows, too.

Next, determine to whom you are selling. This seems obvious: everyone. But you need to be more specific. Here is a list to consider:

- *Department members*—They need to know what the department does, too. Sometimes they know only their own responsibilities, so when someone from the public asks, they do not give the complete answer you hope they would. This need to know should also include all members of a local union or labor group.
- *Policy makers* (mayors, managers, councils, etc.)—They do not always know what you do even though they fund it! They often have so much on their plate that they do not get into the details of your business. (A council member once asked me where the beds were in the fire station even though at the time it was a combination department and had no one staying in the station around the clock!)
- *City managers*—They are generalists with oversight in many different areas. Most city managers do not have a background in the fire service or public safety. They need you to help them learn as much as they can in the shortest amount of time.

- *Other municipal departments*—You will be surprised how little the other departments know about your various services. They are not much different from the general public. As government employees they have a certain amount of influence. If they make certain comments, their position may give them credibility in some people's minds. For example, should they comment that the fire department is overfunded, some people may take that as fact since it came from someone on the "inside."
- *Business community and business leaders*—They have their own perceptions and expectations, often based on their personal history, such as where they grew up, what books they read, and the like. Further, their impressions may come from their only contact with the fire department, in many cases the inspection bureau.
- *Community leaders*—Those in your community who are not elected but wield influence often, too, do not understand the fire service.
- *Citizens at large*—Ultimately, they determine your success.
- *Children*—Through public education contacts, station tours, and any other interactions they take home information to the parents and the rest of the family. They also will eventually grow up and become voters.
- *Others*—These are other agencies outside your community with whom you have interaction (county, state, federal).

Fire departments have continuous opportunities to market through every interaction with the customer (or public). The most effective way is to do the job the best you can and to exceed the expectations of the customer. Every time a job is done well, you gain a supporter, ambassador, and marketer. That support is critical to obtaining the resources you need to do the job you know needs to be done.

What do people really want? That is difficult to answer quantitatively, but you may be able to discern their desires from the letters you receive. Most of the letters received by fire chiefs regarding emergency response contain three elements: appreciation for the quick response, the kindness and empathy of the responders, and the professionalism displayed. There is usually no mention of what the time was exactly, just that fire departments arrive faster than most anyone else. Thus, members must be informed of the importance of a safe, quick response, the attitude they bring on every call, and the need for them to be really good at what they do.

"Customer feedback" on nonemergency calls is similar. In fire prevention issues, rarely is the discussion about the technical aspects of the job. The public cares that you responded in a reasonable amount of time, you were polite, and you acted professionally. Complaints that cite specifics in the fire codes or technical competence are rare. Complaints usually are about perceived poor attitudes or a slow, or no, response to a request. Issues handled quickly with a phone call or personal contact and the proper demeanor will eliminate complaints and often provide support.

Besides examining unsolicited letters and calls, you can find out what your "customers" think the same way any other business would, by asking them! You can develop a simple survey form that provides quick feedback (see Figure 8.2). Many in the service industry already do it— hotels, car rental companies, and restaurants, among others. Check out how they do surveys and see what you can learn from them. Consider sending a postcard with questions you want answered to a random sampling of incidents and contacts, perhaps using a different card for emergency and nonemergency activities. Introduce the survey with a letter and make sure it is simple to complete and that it may be returned at no cost to the individual. Be sensitive to unusual circumstances that may affect usable feedback owing to emotional issues, such as a case involving a fatality occurred. Sensitivity will be required on your part. You may also want to restrict your questions to residents and property owners, in other words, the taxpayers!

Sample Letter and Questions

Dear Resident,

On ________ the Fire Department responded to your emergency call. We hope that everything went as you expected. In an effort to evaluate the performance of the crews that responded and make sure we are always looking toward continual improvement, we are asking your help. Enclosed is a brief survey card. We would appreciate if you would complete it as candidly as you can and return it as soon as possible. Rest assured that your responses will be kept in confidence by me, unless you direct otherwise. Also, although this survey concerns the fire department, if you have any comments regarding any others with whom you may have interacted—for example, the dispatchers, police department, or private ambulance company—feel free to add them and they will be relayed to the appropriate agency. I thank you for your assistance.

Chief ________

Questions

1. Did the fire department arrive in a timely fashion? If not, how long did it take?
2. Did personnel act appropriately for your situation?
3. Were they considerate of your circumstances?
4. Did the firefighters act professionally and skillfully?
5. Do you have any other comments to make?

FIGURE 8.2

People naturally like the fire service, but often don't like "the government." You need to understand this disconnect. Regardless of your own perceptions, fire department personnel are sometimes painted with the same brush as other governmental employees. The department almost always receives its funding through taxes. Since the Boston Tea Party the citizens of this country have demonstrated their reluctance to paying taxes. You need to convince taxpayers that your service is worthwhile (even though it is part of government). You also need to make sure they understand you are an efficient and effective service necessary for the health and welfare of the community. Being **efficient** means able to function without waste, which is exactly what the community wants from its public servants. **Effective** services are ones that produce the desired results, which in this case are not only what the public expects, but what true professionals deliver every time.

efficient
Able to function without waste

effective
Producing the desired results

Certainly, at this point you understand the importance of marketing. Yet, how can it be accomplished without additional resources? You need to get creative: utilize existing resources, seek out low-cost/no-cost methods, and consider all interactions as an opportunity to market the department.

◆ SUGGESTIONS

Here is a sampling of potential marketing opportunities:

- *Newsletters*—Do you or your community put one out to the citizens? If so, use it as best you can and also enlist the help of someone who may know the best way to communicate through that medium. Also, do you have an internal newsletter? That is a way to reach your members with the message they need to hear.

- *Public safety education programs*—Departments get many opportunities to present information regarding fire safety. Add other information to your programs that you feel the audience should hear.
- *Open houses*—Open your doors to the public, to other municipal employees, and others whom you want to show what you do. While they are admiring the bright, shiny trucks and the cleanliness of the station, inform them of the things you do.
- *Charity events*—You can reach influential people.
- *Citizen awards*—Give them out. It helps reward those who do good things and also presents great human interest stories that may cause the media to take notice.
- *Training programs, especially live training scenarios*—People are interested and will stop by if given the invitation.
- *Teach CPR*—Have a cadre of instructors. You help the community and promote your organization at the same time.
- *Parades, large and small*—They provide great visibility.
- *After-emergency programs*—People are interested in hearing what happened in their neighborhood. They will attend to listen. Once you have the audience, you can deliver your message.
- *Link up with other programs*—Burn center programs, police safety, water safety, seatbelts, and child safety seat programs, to name a few, offer additional exposure.
- *Smoke detector giveaways*—You can get donations and return something for free to the taxpayers. Don't forget to use your firefighters to install them, too.
- *Link up with service groups*—Organizations such as Rotary, Optimists, Elks, Lions, PTAs, Boy/Girl Scouts, and so on, have many functions.
- *Link up with private business*—Companies often like to be associated with the "good guys." They can help you because they may have more resources than you. Big companies such as McDonald's, Burger King, K-Mart, Wal-Mart, Dunkin' Brands, Target, ACO, Fireman's Fund, and others have outreach programs. Don't forget your local businesses; those that have been supportive of many community issues also may be willing to help.
- *Take advantage of disasters*—Be prepared to deliver your message when you have everyone's attention.
- *Involve politicians (but stay nonpartisan)*—They also like to be associated with the good guys. Include them in your activities when possible.

These are just a few examples. Use your creativity, and involve others in your department. Your job is to create the program and lead. Do just that.

THE MEDIA

media
Television, newspapers, and radio collectively

The **media**, television, newspapers, and radio collectively, and the people involved in their production, are very powerful and affect your image and ability to market. They determine what people think of you. You need to build a relationship with them and create an atmosphere of cooperation. Always be prepared. You can either be proactive and meet with them before a disaster occurs, or be prepared to work with them when disaster strikes and they are interested in your story (see Figure 8.3).

Much has been written regarding media relations. They are not much different from relationships with others, and building them takes time and a commitment to do the job. Know your media outlets and who covers your area in print, radio, television, and perhaps cable. All are important. They all contribute to the delivery of news and information in your community in one way or another (see Figure 8.4). If you have an

FIGURE 8.3 ◆ Television cameras will always be around. Fire chiefs need to work on their media relations.

ators offer opposing views on Blue Cross reform

Library to 'go live' with automation system

DECA students in the market for international success

Officer Herrmann honored by MADD

Family Fun Walk will benefit senior services

Tee time set for Junior Optimist golf tourney

A man doesn't go to the doctor when he's healthy.

Beaumont

4.95%

Credit Union ONE

Credit Union ONE

FIGURE 8.4 ◆ The print media are as important as television.

individual in the department who has an aptitude for media relations, assign the task to him or her and support the individual. Do what you can to help that person learn the skills they will need. Provide resources, especially people and places to seek out information. If you do not have such an individual, you will need to do it yourself, because it is an important part of the job.

◆ PREPARATION

Although some have a natural affinity for the media, most people could benefit from training. This can be obtained through formal programs—workshops, seminars, and other educational opportunities—or maybe through a local university or community college. Don't be afraid to approach local media and ask for help. Often, they are happy to share their experience. They know that if they help you, they will get better treatment when they need to cover a story in your community. You can also learn from your network. Are there others who seem to get good coverage? If so, ask them for advice.

Whenever an emergency will be news, be prepared. Your time will come, whether a serious fire, accident, or other newsworthy incident. When it happens, formulate your thoughts, and anticipate the questions to be asked. Former Secretary of State Henry Kissinger was once quoted while starting a news conference, "I hope you have the questions for my answers." You, like the secretary, have a message to deliver and must refuse to allow yourself to be sidetracked from the critical issues. This approach should be considered in many circumstances, although it will not always be beneficial to you or your organization. Most likely, the questions will involve who, what, where, when, why, and how. Some departments have created information sheets that can help provide answers to the questions you will be asked (see Figure 8.5). If you are unable to think of your own questions, ask someone else. In most cases the questions asked will not be a big surprise. Be calm, be professional. If you have a good relationship with the media, you may even ask before the cameras start rolling about the intent and direction of the interview. Most of the time, reporters will not tell you specific questions but may give you some general information. Cooperate as much as you can. If you are perceived to be helpful, you are more likely to get fairer treatment and maybe even the opportunity to get your message in the story.

People will judge your organization by your presentation. The image you present will tell others much about your organization. In addition, when the questions are asked, try to be as personable and professional in your answers as possible. Always be honest about what you can say. If you are not able to answer a question, do not make up an answer or avoid the question; say you cannot answer or do not know the answer. Finally, practice when you get the chance. The more you do it, the more relaxed you will be and the more competent you will appear.

◆ CUSTOMER SERVICE

customer service
Work done for clients as a job, duty, or favor

Customer service has become a very common expression and refers to the work done for clients as a job, duty, or favor. From the very beginning fire departments have been providing great customer service, so the concept is not new. However, today the level of service provided needs to be enhanced to develop customer "loyalty;" that is,

SAMPLE MEDIA WORKSHEET

Spokesperson ______________________________ Date ______________
____

Type of incident ______________________________

Location ______________________________

Total number of units involved: Engines __________ Ladders __________
Medics __________ Squads __________

Total number of firefighters ________ Mutual-Aid companies?
Type: ______________________ Number: ________

From where:
____
____

Response time __________________ Time of dispatch ______________
Under control __________________

Fire under control in __________________ minutes or
______________________________ hours

Cause of fire
____
____

Total number of injuries ________ Civilian ______________________
Fire Personnel ________________

Transported to
____
____

Types of apparent injuries
____
_____

Severity (circle): Severe Moderate Minor

Total number of fatalities __________ Transported to
____

Brief synopsis of first arriving unit(s)—What was encountered and what action occurred?

____
____
____
____
____
____
____
____

Special hazards encountered—toxic material, flammable liquids

____
____

(*continued*)

Special accomplishments—rescues, salvage work, etc.

Dollar loss estimate
$

If unknown, estimate (circle one) Heavy Moderate Minor Unknown

Observations made by Fire Prevention Division

Media contacted by PIO

Comments: The incident is currently under investigation. Once the investigation is complete, details will be available.

FIGURE 8.5

it is necessary not only to take care of the problem but to make sure the customers are so satisfied that they are moved to tell others and offer future support. Remember, all your personnel must be nice to everyone all the time!

Great customer service is not difficult to provide. Figure out what people want and give it to them and add to it. Go the extra mile. People's needs and wants are the same when you discuss the basics of customer service. Understand that these cannot be met all the time, but you must try. Be proactive and empower your people to use common sense and "do the right thing." Of course, make sure you provide the necessary training to your personnel, as not everyone has the same degree of common sense.

Consider what you would do if you or one of your loved ones requested the service. What would you want? Perhaps you would like perfect service or something that makes your day just a little better. Remember, the person who called is having a bad day. He or she wants you (or government) to fix it. You make it better by just responding, usually very quickly. Everything else enhances the service. Create the culture that encourages extra effort. Your personnel will respond.

As an example, consider what your personnel could do to provide service on a very basic incident. In many situations following an injury accident, the individual injured, who is often alone, would like to contact family. If it is possible, provide a cell phone to make the call possible. This may seem like a little thing but can be huge to someone involved in an accident. This is not only good service but the right thing to do.

There are differing views on what the public wants. In general terms, they want prompt service, caring and competent people to respond, and to know that the taxes that they pay are being used wisely. This is customer service, which most of the time is exactly what fire departments do. Even though a department may respond to thousands of calls each year, individual people will make an assessment based on a single contact.

In the past, the concept of customer service was passed on informally. Depending on who had influence, on the commitment of the company officer, and on the culture of the organization, personnel learned what needed to be done to be successful. The degree of participation varied within the department, from shift to shift and station to station. Today, if you want improvement and consistency, you need to provide training and policies. The training makes sure that everyone hears the same message and the policies document the desired behavior. Together they help create the positive example needed to establish the culture of quality customer service. As with any program, you send a message to your organization by the amount of time and effort you commit to a program. If you think customer service is important, commit the resources to enhancing the abilities of your personnel.

You as fire chief need to market not only the department but yourself. To some extent, you live in a fishbowl. All your actions are being evaluated every day by someone. Your image is important for the organization and you as an individual. Thus, it helps to be the positive poster child for the fire department. To do that you need to do the following:

- *Look professional.* Remember your appearance at all times. Be conscious of dress, posture, body language, and facial expressions.
- *Act professionally.* You are held to a higher standard whether you like it or not. Act appropriately at all times whether or not you are in the public eye. Anything other than professional behavior by the fire chief will hurt both the department and the chief.
- *Be a good example.* Even though it may not be possible to be the expert in everything in the emergency service, learn what you can so your knowledge base is respected. Never stop learning. Know your business. People expect it.
- *Be concerned.* Have a passion for your work.
- *Always treat people as individuals.* Avoid using the excuse "it's policy" when dealing with individuals.
- *Be correct or seek out the correct answer.* Protect your integrity.

◆ SUMMARY

It is vital to recognize the role of marketing, media relations, and customer service in the overall objective of the fire department. You are competing for people's money. They want to know what they are paying for and want to know that they are getting value. They will not take the time to do an in-depth review of your services. They will give you a limited opportunity to provide them the information they need to make their decisions. That is the world in which you live and you must adjust your thinking and your way of doing business to fit into their world.

ACTIVITIES

1. Identify the media outlets in your community and develop a plan to improve relationships.
2. Develop a "pre-plan" information sheet that can be used to anticipate the questions likely to be asked by the media during or shortly after an event.
3. Survey your department members to ascertain their understanding of customer service. From the survey, develop a presentation to the organization that will communicate to all the preferred delivery of customer service.
4. Develop a list of five fire departments that you perceive provide good customer service. Identify the elements that promote this image.
5. Identify the image you would like for your department. Perform a survey of your community. Do your perceptions match those of the public? If not, develop a plan to change perceptions to more closely match desirable ones.
6. List the various groups in your community that would benefit from increased marketing by your fire department. Develop a plan to deliver your message.

◆ CASE STUDY

ADDED VALUE TO AN INCIDENT: CARBON MONOXIDE ALARM

Your department receives notice from the dispatch center of a call for a carbon monoxide alarm activated. The dispatcher contacts the on-duty shift commander. It is 3:00 a.m. The shift commander, who was awakened, asks if the caller reported any medical symptoms. The dispatcher responds no. The shift commander instructs the dispatcher to tell the caller to unplug the unit and call back in the morning. The next day, after shift change, the new shift commander reviews the calls and relays to you the events of the previous evening.

What will you do?

Finance, Budgeting, and Purchasing

CHAPTER

KEY TERMS

budget, p. 105
discretionary, p. 112
fees, p. 106
finance, p. 105
mandatory, p. 111
purchasing, p. 105
taxes, p. 106

Money—you can't provide a quality fire service without it. Somewhere in the job description for fire chief, it says that the fire chief is in charge of budgeting and spending on behalf of the fire department. What it doesn't say is that generally there are some rules to be followed and some restrictions placed on the fire chief regarding the types of things he or she can do. Thus, it is extremely important for you as the fire chief to know about finance, budgeting, and purchasing. You don't need to be a CPA or a math wizard, but you must know how the system works, how you fit into the system, and what policies and procedures apply. The following definitions apply to this discussion: **finance** means the business of managing the monetary resources of the community; a **budget** is an estimate of the income and expenses of the community; and **purchasing** means buying something using money.

finance
The business of managing the monetary resources of the community

budget
An estimate of the income and expenses of the community

purchasing
Buying something using money

You also need to know what your role is in the process. As a general rule, you do not authorize, appropriate, or apportion the funding, you request it. The amount of money set aside to fund your organization is a policy decision set by the policy makers in your community, who may be the mayor, city council, county board, or fire board. Your job is to spend the funds wisely while trying to attain the goals of your organization. Although you do not set the amount, you do make the request and can influence the decision through your relationships and politics (which have been covered in a previous chapter). You may get something because the policy makers like it, because they tend to fund the programs they prefer, not necessarily what you prefer. Politics and budgeting, at least from the appropriations standpoint, are definitely interrelated. Nevertheless, you must follow a set of rules and understand the basics of finance, budgeting, and purchasing so that you can operate within the guidelines of your organization. See Figure 9.1.

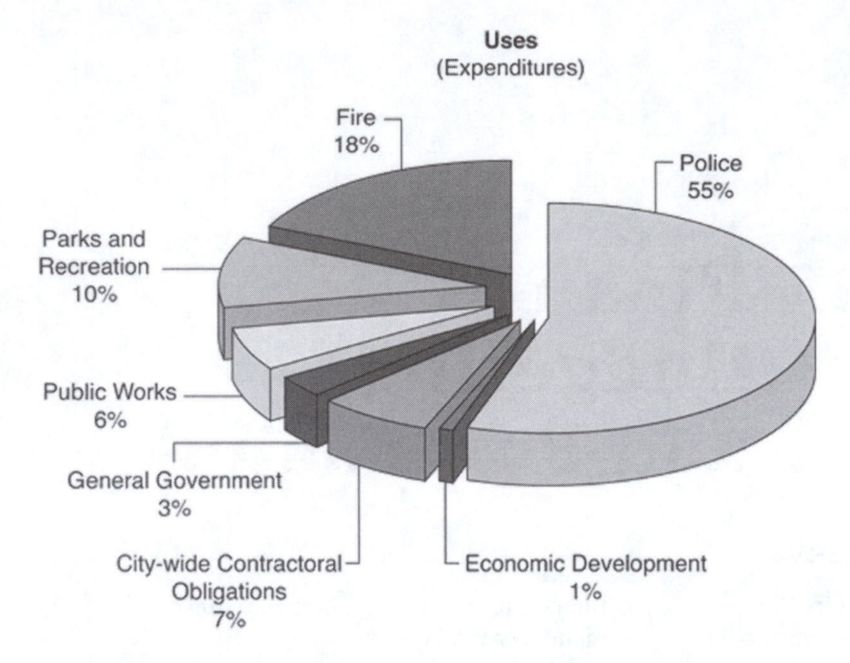

FIGURE 9.1 ◆ Be aware that the policy makers have many demands for spending taxpayer dollars.

◆ DOING THE MATH

There are two ways to look at the finance and budgeting portion of your job. One is from a management perspective: your job is to balance the budget. This is a simple math problem. You are given a certain amount of funding, so you account for all the things in your organization, assign a cost to them, and make sure that this cost does not exceed the amount of money you have been given. Making the income and expenses match can be done relatively easily. Of course, this does not take into consideration the outside influences, emotions, and factors not controlled by the fire chief. Another way of looking at the budget is from a leadership perspective: a true leader approaches finances with the intent of improving services by earning the trust of the public while balancing the budget. The intent is to provide value for the taxes that are paid by continually searching for the best ways to deliver the results the citizens want at the price they are willing to pay.

As you get started in the process, you need to understand about taxes, fees, and other sources of income (see Figure 9.2). Most fire departments are supported by taxes generated by homeowners and businesses. **Taxes** are the amount of money levied by the government on its citizens and used to operate the government, such as income or sales taxes. Other funds may be generated through **fees**, that is, payments for some forms of professional service (EMS billing, inspection fees, false alarm fines, cost recovery for incidents, etc.). You need to know all your sources and whether they are dedicated strictly to the fire department. In some cases, taxes (or other fees) are earmarked for the fire department through a vote of the people

taxes
The amount of money levied by the government on its citizens and used to operate the government

fees
Payments for some forms of professional service

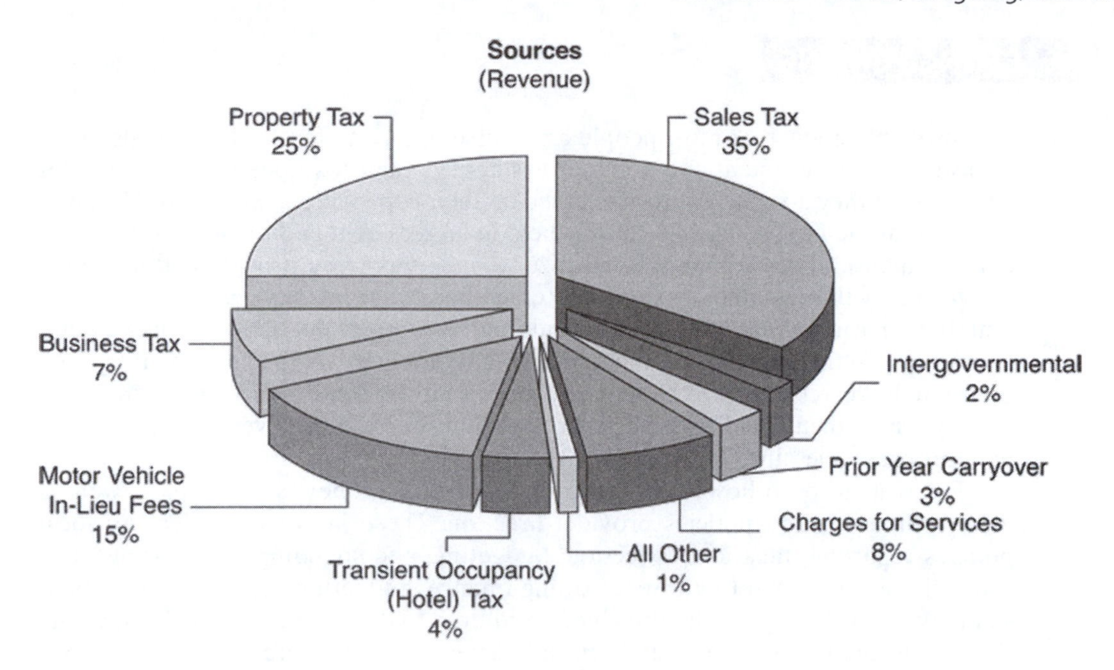

FIGURE 9.2 ◆ It is important to know all the sources of income.

or by local legislation—a dedicated fund. This tax is collected for purposes of operating your organization. Even though it may be dedicated directly toward your organization, there may be some restrictions. Check out what you are allowed to do. Sometimes, it may create a more confusing financial situation for you. For example, some communities that collect direct taxes for the operation of their fire department will then charge the fire department for other internal services such as budgeting, vehicle repair, and station maintenance. Some policy makers believe that the community will not support certain services with their tax money but will fund the fire department (or police or public safety in general). Making the fire department pay for their support services effectively transfers money to other community departments without having to seek the funds directly from the taxpayers. Hence, this is another reason to understand how the budget game is played in your community.

Often, taxes collected go into a general fund to support all local services. The policy makers then appropriate the funds as they see fit among police, fire, parks, public services, general government, and others. Some communities, maybe even most, have combined sources of income, some of which may be committed exclusively to the fire department. Some in the fire service may not always view this policy as fair, because they feel that funds generated by the fire department should stay with the fire department. The counterargument is that other municipal departments also generate funds, such as building permit fees, but don't always get to keep them. Regardless of your personal opinion, the policy makers will make the determination and it is up to you to understand the process and applicable rules, policies, and procedures.

◆ THE PROCESS OF GETTING FUNDING

You probably know that most people are averse to paying taxes. They are also very cynical about government and don't always believe their tax dollars are being spent wisely. What they do want are results based on their expectations, not yours. The good news is that the fire service is generally held in high esteem and recognized as a necessary expense of government. Recognize what services you provide and how they are valued by the customer (taxpayer). Remember, it is what they want that is important, but bear in mind that budgeting is ultimately a policy decision. To some extent, people choose their level of government by what they are willing to pay in taxes. They occasionally direct funds to the department but mostly send their taxes to the community, which then decides, through a representative form of government, how best to appropriate the funds (see Figure 9.3).

Rules must be followed in spending taxpayers' money; state laws as well as local ordinances and policies provide direction. These laws usually lead to local policies regarding taxation, collection, budgeting, and spending (purchasing). Because there is so much to know regarding finance, budgeting, and purchasing (and so many other things that the fire chief is required to do), you don't always have the time to do the necessary research on the subject. This is where your local finance personnel (finance director, budget director, comptroller, etc.) can be very helpful. They operate with budget and funding issues every single day. They are the experts

FIGURE 9.3 ◆ There is a lot of competition for budget dollars. You need to influence the policy makers.

in the field, so build a relationship with them. Take the time to sit down to learn what you can from them as you begin the process. Even if you think you have a good understanding of finance, things can change from year to year. You also budget annually, not daily. It is extremely helpful to have the insight of those who deal on a daily basis with the financial issues of your community. Refresher training is good, and it also helps build that very important relationship with all the finance people.

There are many books regarding public finance and budgeting that can provide the details you may need to feel comfortable with the process and system being used. You can check with your local bookstore, do an Internet search, or ask some of your contacts in the finance world. Books are continually being published and processes adjusted or amended. You may also check with the Government Finance Officers Association for additional resources. As fire chief you need to expand your horizons and learn where to look for the most current information. You should understand that there are various types of budgets, some more "trendy" than others.

This unit will not explore your particular system to the detail you need to know. The purpose here is to raise your level of awareness, point you in different directions where resources may be available, and keep you out of trouble, all the while recognizing that you have relatively little time to do this in light of all your other job responsibilities. You also need to understand that your budget is a reflection of your core values and a statement of what you perceive to be important. It is also a planning document. Do not view it as just rows and columns.

BUDGET SYSTEMS

Initially, you should know the budgeting style of your organization. The most common form is the line-item budget. This is a system that identifies specific items and services to be purchased with their corresponding (estimated) cost. It can be very detailed, identifying relatively small purchases (such as very specific fire tools), or a bit more general, allowing for some discretion once the budget is approved (see Figure 9.4). Other budget systems come into vogue from time to time, such as zero-based budgets, performance-based budgets, or budgeting for outcomes. Once you know the style being used, try to learn as much about the system as you can. It not only helps you with the numbers part of the process but can be of value politically. Bear in mind that any changes in government leadership and/or finance personnel may change the type of budget used.

In general, and regardless of the budget system being utilized, there are usually two categories of budgets: *operations* and *capital*. The operations budget deals with the day-to-day functions of the organization, including personnel (salary, benefits, etc.), supplies, services (professional and contractual), and generally smaller fixed assets. Capital budgets are established to purchase the big-ticket items needed to operate the organization, including fire trucks, fire stations, costly equipment designed to last for years, and other items that generally are not purchased annually and have a significant cost. Check into your local policies regarding the fixed dollar amount that qualifies a purchase as a capital expense, as this varies by locality.

SAMPLE LINE-ITEM FIRE OPERATING BUDGET

DEPARTMENT NUMBER:

Acct. No.	Category and Line Item	Actual	Actual	Budgeted
(702)	**SALARIES & WAGES**			
010	Administrative & Clerical	4,043,956	4,572,038	4,528,056
038	Part-time	2,707	0	2,000
042	Holiday Pay	96,430	104,762	110,324
106	Sick & Vacation	73,236	104,998	87,351
112	Overtime	353,402	364,969	350,100
200	Social Security	348,084	393,853	399,679
250	Blue Cross/Optical/Dental	471,756	590,421	659,874
275	Life Insurance	6,626	7,142	8,236
305	Pension—DB	775,155	854,284	864,241
325	Longevity	105,764	117,204	131,647
350	Workers Compensation	87,531	83,012	75,500
	Category Total	6,364,647	7,192,683	7,217,008
(740)	**OPERATING SUPPLIES**			
001	Gas & Oil	63,471	64,037	68,500
002	Books & Subscriptions	5,993	5,816	6,874
008	Supplies	157,837	120,176	97,080
011	Medical Supplies	41,617	53,216	41,500
019	Uniforms	28,625	40,266	35,500
020	Protective Clothing	31,164	223,502	7,000
040	Miscellaneous	27,598	26,005	22,250
075	Fire Equipment Repair Parts	16,321	22,278	11,250
076	Fire Prevention Materials	5,058	2,876	5,950
	Category Total	377,684	558,172	295,904
(801)	**PROFESSIONAL & CONTRACTUAL**			
001	Conferences & Workshops	2,375	2,725	5,330
002	Memberships & Licenses	8,766	9,728	13,730
005	Fleet Insurance	35,000	35,000	31,800
006	Vehicle Maintenance	31,164	45,996	35,000
007	Office Equipment Maintenance	5,557	7,128	7,125
008	Vehicle Refurbishment	(7,586)	0	1,425
009	Consultants	44,422	54,854	50,000
013	Education and Training	21,416	39,896	47,330
016	Phone Expense	20,734	19,248	21,000
023	Data Processing	27,270	33,279	29,626
025	Utilities	133,543	144,271	165,000
026	Physical Examinations	28,830	23,952	44,495
027	Radio Maintenance	8,788	8,827	9,300
029	Building Maintenance	40,372	34,241	51,652
032	Fire Equipment Maintenance	0	0	0
	Category Total	400,651	459,145	512,813

(970)	**CAPITAL OUTLAY**			
001	Station Furnishings	0	3,110	4,200
002	Office Equipment	0	0	72,000
007	Equipment	13,771	5,290	246,740
015	Vehicles	24,069	84,039	50,000
019	Radio/Communications	3,488	20,038	25,000
036	Building Improvements	0	0	13,760
075	Training Equipment	0	0	75,944
077	Hazardous Materials Equipment	2,671	0	1,325
	Category Total	43,999	112,477	488,969
	DEPARTMENT TOTAL	**7,186,981**	**8,322,477**	**8,514,694**

FIGURE 9.4

◆ PREPARING THE BUDGET

In the normal course of most budgeting processes, you begin with what you had. Even with program planning and zero-based budgeting, the reality is that you are not starting with a blank sheet of paper. Much of what is done is carried over. One-time purchase items are eliminated and new ones are added. However, significant changes most likely will need extra effort to justify. You must have the cost of the personnel, items, or services (amount to be budgeted) and a source of revenue for the added expense. This is where you need to do the necessary homework and research to prepare for the coming scrutiny that will often determine whether the presented budget is accepted and adopted. You certainly need to know all the direct and indirect costs and then must be able to articulate the benefits to the department and to the community so that the policy makers can make a decision, in your favor, it is hoped (see Figure 9.5).

◆ MANDATORY VERSUS DISCRETIONARY SPENDING

Within your budget are many **mandatory** expenses, those that are implied to be required by ordinance or by the people. Once you get into the process, you will understand that the fire chief has very little discretion over the expenses within the budget. First and foremost, in most departments with the possible exception of volunteer departments, personnel costs can account for anywhere from 80 percent to 97 percent of the operating budget. This cost is the major reason that many politicians are resistant to adding personnel. Once a firefighter is hired, the pressure to keep the personnel can be great, regardless of any changes in a community's ability to pay. The costs for personnel continually rise, sometimes faster than communities raise funding. Mostly for political reasons, reductions in staffing levels are very difficult to implement. As a result, this expense rarely goes away.

mandatory
Required by ordinance or by the people

FIGURE 9.5 ◆ Elected officials oversee spending. Know their role.

The wages are only one component of this expense; benefits have become very expensive as well. The benefits are viewed as an entitlement, one that may be perceived to be out of the control of the policy makers. Further, as the fire service is a round-the-clock operation, it generally takes four to five firefighters to staff one position at all times. This includes the three to four shifts plus coverage for vacations, training, sick/personal leave, and any other paid time off. The other part of this challenge is that fire and emergency services are very labor intensive when needed. Hence, budgeting is as much about politics and policy as it is about funding.

Of course, there is always the possibility that staffing levels can be changed, usually as a result of the rapid growth or decline of a community. Whatever the reason, hiring additional firefighters or laying off personnel will create an extremely difficult process for everyone in the organization and will require appropriate justification. These issues contribute to the understaffing of most organizations. Most competent managers (the fire chiefs' supervisors) tend to understaff because of the cost of firefighters, not because the managers don't believe they are necessary. You as the fire chief must understand the challenges this situation can present in the day-to-day operation. If there is a minimum staffing requirement and there are not enough firefighters on staff, overtime will be necessary. The budget to pay the overtime costs will be closely scrutinized, so you must know this part of the budget as well as or better than any other part. You must understand the difference between mandatory (statutory or regulated) versus discretionary overtime. **Discretionary** means that you have the freedom to make decisions based on the circumstances. A labor agreement may mandate

discretionary
Freedom to make decisions based on the circumstances

the staffing and subsequent requirement to pay personnel overtime to meet contract provisions. The Fair Labor Standards Act (FLSA) also has requirements regarding hours worked and overtime obligations. This is a complicated law and requires legal advice to apply it. Sick time, vacations, other leave (bereavement, required training, etc.), and disability time off due to on-the-job injuries can create substantial needs for overtime. Even though the overtime issue may be outside your control, you may be expected to control it. These are the sorts of things that must be addressed and openly discussed in the budget review process. Further, understand that hiring additional personnel will not make overtime go away. All employees have leave time, and their schedule will not automatically fill the spots that created the overtime. Always be upfront about overtime costs.

In addition to your personnel costs, other nondiscretionary items within the budget include your utility payments, fuel for your apparatus, necessary supplies to run the organization, and a host of similar items. These are the types of services and products that must be paid for and are absolutely necessary to continue operating your organization. For example, you could not run your business without telephones, and you must pay the monthly bills to the phone company. Hence, you really have no choice in this particular area of your budget, but you can shop for the best prices and/or services whenever possible. For example, there is much more choice in phone service today than formerly. Cost comparisons on a regular basis can save you money.

You can easily identify those areas in your city and fire department budget that are mandated. In most cases, you will find that, unless it is a true zero-based budgeting process or a budgeting for results process, you have very little that is new from year to year. Because the fire service is a labor-intensive business, personnel are the largest cost. Starting from zero every year and justifying your staffing is not practical and is rarely required. The other components of the department usually have their supporters, both inside and outside the department. They also are the relatively less costly items within the budget. Therefore, it is very easy to keep programs going and difficult to stop ineffective ones. The traditions of the fire service are hard to break, even if they don't make logical sense. Remember, the budgeting process is more of an exercise in making the numbers fit than in developing different types of programs in your organization.

It is important to emphasize the need to build relationships with your finance department. They are the true experts in your community. They will know the ins and outs and ultimately will have control of your spending within certain parameters. They also can give you tips on how to save money and streamline some of your processes. Take the time before the budget preparation season to understand not only the budget process but also the people in your organization who will be working on it. Cultivate relationships not only with the elected officials but also with the bureaucrats directly in charge of the funds.

◆ BUDGET ADOPTION

After you have been provided direction with respect to available funds you begin the math problem of appropriating the anticipated income to spend on the items you need (personnel, equipment, supplies, and services) and complete a draft budget.

Your community will have a process for adopting the budget, which can include an internal review (city manager, mayor, or staff along with the budget and finance people), review by the policy makers (elected officials), and a public hearing (probably mandated). This process can involve various community study sessions culminating with a formal approval by the policy setting board.

Often, completing your budget is the easy part. It is the review that requires most of your energy. There is most likely a formal and an informal process, at each step of which you are asked to convince someone or some group that what you are requesting is necessary for the good of the community. It is the "sales" part of your job. You get past the point of the simple math of the budget and get into the political realities of getting your programs approved. Preparation is essential: do your homework, and know not only the numbers and details of the budget but also the players involved. Through your relationships, you should be aware of the types of budget issues that are important. Review the budget with others in your organization and ask them to question various aspects of it. This will help you anticipate questions and prepare answers. If you can't answer them in private, you won't be successful on the public stage. Also be aware of any public concerns. Your preparation can determine your success.

◆ SPENDING

Once the budget is approved, you are then able to spend the money. Obviously, the major portion of the budget is already spoken for, that is, your personnel costs. Generally, those funds are set aside, and things such as payroll, insurance, and pensions are paid for directly; control of that spending is restricted. It is important to note that those particular items have the greatest impact on your budget, not only because they are the largest percentage but because they currently are rising at a rate outpacing inflation and may be increasing faster than the existing sources can generate income. As this trend continues, the fire chief is asked to make cuts in other areas. Today, health care and pensions are placing a tremendous strain on communities trying to balance their budgets. Also, inadequate staffing stresses the overtime budget. Regardless of your beliefs, you can spend only what you have been given.

For other than those fixed expenses, there is most likely a purchasing policy within your organization. Know what that policy is. Failure to understand and follow the policy can land you in trouble and also create poor relations with those responsible for purchasing. It will also delay your purchase. As with most parts of your job, you do not need to know every detail of the process but must know the important points. Any of your staff involved in the process must also know them. The policy will include several specifics such as spending limits for individuals, processes for soliciting formal and informal bids, and the drafting specifications to be considered. Within the policy, there are general spending limits. There is an amount that general employees (fire officials below the rank of chief) and the chief can spend. A higher amount will require the authorization of a higher authority, such as a budget director or comptroller. Yet another amount will require the approval of the senior executive, for example, the mayor or city manager. Finally, there will be an established spending amount that will require the approval of the governing body.

Like it or not, you must follow these rules. Remember—request, approval, and purchase, in that order. Even if there is a written policy, you need to consult with the purchasing personnel in your organization. They are the ones who know the policy inside and out. Further, they can help you negotiate through the process when you need to make a purchase. Conversely, they can also slow down the process if you don't follow the rules. It is suggested that you involve the purchasing personnel early in the process, especially when quotes and formal bids are required. They will appreciate the inclusion and help you negotiate through the maze of the policy. You can get what you want and need following the process. Building relationships ahead of time is important. Do not create extra work for yourself by neglecting the people or policy, even if you don't agree with them or it. Accept your system and work with it.

You may or may not know the lingo of purchasing. Like many in government, the people in purchasing have acronyms and other terms that are specific to their process (not unlike some of the unique language used in the fire service). Learn what you can from the experts. You should know the difference among quotes, bids, formal bids (with official bid openings), specifications, and RFPs (requests for proposals). Again, remember that you may have limited opportunities to work within the purchasing policies. Your purchasing department uses them daily with many departments, so they know them better than you. Consider them a valuable resource and have them assist you whenever possible.

◆ SPENDING LIMITS

Depending on the spending limits established, you will need approval from others when the cost of your purchase reaches a certain dollar amount. At this point, you need to be prepared, for your request will be scrutinized. It sometimes helps to study how the policy makers treat others who present purchasing approvals. How do they question the police chief or other department heads? Often, they know little about your business and ask few questions when dealing within your area of expertise. They will ask questions in areas they know. Depending on their background, they may be extremely knowledgeable (or think they are). For example, it is rare to get questioned about specialized equipment specific to the fire department because few on the elected board know enough to ask. They will not want to be embarrassed by asking a "dumb" question. They will ask generic questions such as why it is necessary, whether funds are available, and how the widget you are buying will help with service. When the purchase is more in the mainstream, and members of your board think they know about it (or it is something in their line of business), be prepared for more detailed questions. For example, it sometimes seems as though everyone is a computer expert. Whether they are or not is irrelevant. Expect that someone will raise issues during a public session to let everyone know they are "watching the purse strings." A cynic may say it is a good opportunity for a politician to get some acknowledgement while questioning something they possibly know a little about. You should understand how important it is to be very well prepared so you are not embarrassed in public, and you get the funding you are requesting. Trust will also be a factor. If they trust you and your boss, you will have a smoother process. Remember, one problem in this area will make all subsequent purchases subject to more scrutiny.

If you are with a large organization, you may be purchasing things in such a volume that you get a great price. If you are like the majority of fire departments, you are midsize with average buying clout. To save money, consider forming group-purchasing consortiums to take advantage of larger orders. Fire departments purchase the same things. By joining forces, you may increase the volume to a point at which the vendor offers better pricing. The larger the consortium, the bigger the savings. Of course, this is as much a political issue as a cost-saving one. Your relationships with neighboring departments will determine your success as much as or more than the actual revenue saved!

One final area to make sure you are well versed in is that of emergency expenditures. On occasion, critical equipment or apparatus breaks and you need an instant repair or replacement to continue service. In the event of a major natural or man-made disaster, you need to keep your operations functioning. What rules will you need to follow to respond to the public's call for help and still comply with your organization's procedures? At the time of the need is not when you should learn how to make these purchases nor try to determine where the money will come from. It may be reassuring to think that whatever you need to spend will be covered by the governing body or via state or federal repayment, but during the middle of an emergency is not the time to get the answer.

◆ SUMMARY

Knowledge of the finance, budgeting, and purchasing methods and policies of your community is vital to your success as the fire chief. Nothing is done without money. Money cannot be collected or spent without following policy. Others outside the department are usually a part of budgeting and purchasing. Do all that you can to learn the systems being used to budget and purchase. Know the players involved in the process, and build your relationships with them. Accept this role as one of your most important.

Activities

1. Identify the mandatory and discretionary elements in your fire department budget.
2. Identify the funding sources of your fire department. Consider taxes, fees, grants and any other sources. Determine which are dedicated to the fire department and which come from the general fund.
3. Obtain a copy of the purchasing policies of your community. Determine the various dollar amounts requiring different levels of authority to approve purchases.
4. Develop a strategy to create a group-purchasing consortium in your community with neighboring departments. What could be purchased and by whom? Who would be involved in your discussions?
5. Assume your staffing levels are inadequate. Develop a proposal to increase staffing. Determine the added staff needed and the sources of income that could be used to fund the increase. Consider any political ramifications involved and suggest strategies to address these concerns.

VIOLATION OF PURCHASING POLICY: SPLITTING THE PURCHASE ORDER

Your city has a purchasing policy that requires a purchase order for any item or service over $500, including the purchase of multiple similar items costing less than $500 if the total exceeds the amount specified in the policy. One of your members was instructed to purchase four replacement helmets that were recently damaged. You have already given helmets to the members who needed them from your stockroom and need the replacements to keep your stock at the level you feel appropriate. The cost of the helmets is approximately $200 each. Your member calls your supplier and requests the four helmets and asks that two invoices be submitted, splitting the order in half to keep the amount under the amount that would require a purchase order. When the invoices arrive, you receive a call from the director of the purchasing department asking you to explain what happened.

What will you do?

CHAPTER 10 Ethics

KEY TERMS

beliefs, p. 123
ethics, p. 118
morals, p. 118
values, p. 123

ethics
The study of moral standards and how they affect conduct

morals
Involving issues of right and wrong

Ethics has become a very important topic in government and therefore in the fire and emergency services. Simply defined, **ethics** is the study of moral standards and how they affect conduct. **Morals** involve the issues of right and wrong. Because you should be concerned with your conduct and that of your personnel, as well as with what is right and wrong, an understanding of ethics is essential. To some extent, there is an element of ethics in every decision that is made. As an agency supported by tax dollars, fire departments must operate within specific guidelines. These guidelines (some of which have the power of law and are very specific) were created in response to concerns regarding government practices and the need to try to establish trust. Government is under a microscope. The general population has a higher level of expectation of their elected and appointed officials, whether on the federal, state, county, or local level. There are checks and balances for spending tax dollars (and some would argue that these are unreasonable and forestall change and efficiency). People want to trust that their hard-earned tax dollars are not being wasted or abused and that the conduct of their public officials is above reproach. These officials include the members of the fire department.

It has been said that a "public office is a public trust." This can be interpreted to mean that those in a public position, especially those in high office, are held to a higher standard of behavior and performance. Fire officials are expected to behave in accordance with accepted practices. Integrity is a critical component of being a fire chief; fire chiefs think it is important and so do their bosses. People trust the fire service, and the head of the organization sets the tone, so this responsibility must be taken seriously. You need to accept that you will be held to a higher standard and operate and act accordingly.

Failure to recognize ethical issues and act appropriately can get you as the fire chief in trouble and even result in loss of your job. You are more likely to encounter ethical concerns than operational issues, and the former can embarrass the organization and you directly. Conversely, ethics offers the fire service a competitive edge in

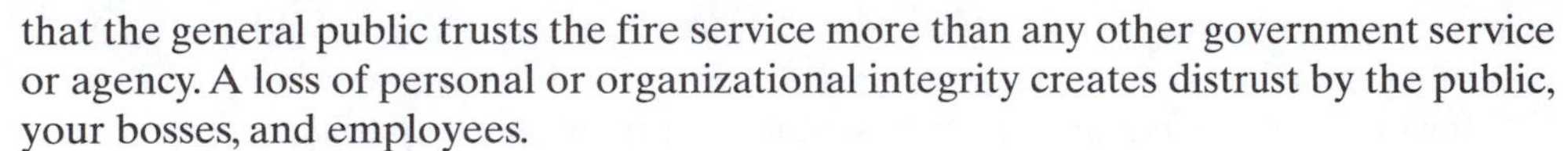

that the general public trusts the fire service more than any other government service or agency. A loss of personal or organizational integrity creates distrust by the public, your bosses, and employees.

Ethics is difficult to define, harder to teach, and nearly impossible to measure. The term *ethics* comes from the Greek word "ethos" which means "customs, conduct, or character." The importance of ethics and morals to a chief of a fire department cannot be overstated. Fire chiefs are more likely to get into trouble for ethical reasons than for any other reason—more than mistakes on the fire ground, lack of leadership, or poor managerial practices.

Ethics are about right or wrong, people's perceptions, and individual beliefs. Ethics may or may not be related to legal issues. Because issues related to ethics are subject to interpretation and the world is so diverse, there is much gray area and very few circumstances that produce black-and-white answers. It is difficult to produce a definitive guide in the area of ethics because of the differences among individuals and organizations. People, history, extenuating circumstances, and established organizational cultures make specific directions and answers to ethical questions difficult to establish.

◆ WHY STUDY ETHICS?

There are very important reasons for studying ethics and preparing for the inevitable questions that will arise in your organization. First, you need an understanding of your community's and department's ethics to perform your duties within the expected norms of your organization and to protect yourself. For example, does your organization have a gift policy? Are you allowed to accept gifts? If not, have you ever received a free lunch, baseball cap, or other item of minimal cost? If so, then does the issue become one of value? Often, the initial answer is, no gifts. Further probing finds that there are exceptions. You need to know the limits in your organization. If you are the one setting the limits, make sure you communicate them to all in your organization.

The second reason to study ethics is to understand how the rest of your organization may think. If everyone thought the same way, there would be no need to establish rules, regulations, policies, and procedures, and decisions would be easy. Although general principles and morals guide society, each individual and each organization have their own variations, practices, and interpretations. Organizations develop theirs over time through their history and culture.

Individual ethics are established as people mature based on many factors, some of which change over time. In fact, the ethics of individuals can change with various events in their lives, and an individual's ethics may or may not match those of the organization. If they don't, the individual will need to change, even if he or she doesn't agree. For example, disciplinary action is designed to change behavior, which on some occasions may be ethical behavior. Most likely the discipline will not change the individual's belief system, but it will change the behavior so it coincides with that of the organization.

To play by the rules and protect yourself, you need to educate yourself about the standards established in your community and department. You don't want to do something foolish because of a lack of understanding of the basic established principles. Ignorance of the law or rule will not give you a free pass. You also need to align your thinking with that of the organization. For example, some organizations have a strict no-gift policy, whereas others establish a dollar limit. You would not want to accept a gift and later find that your organization prohibits your action.

It can also happen that an agency's ethics do not measure up to an accepted standard. For example, some departments have taken lengthy and expensive trips to inspect the factories where their apparatus was being built. By coincidence the factories were in warm-weather locations. This is not to say that factory inspections are not important, but sending an inappropriate number of people for an excessive length of time may create an ethical dilemma. In such a case, you may wish to pursue changing the organization standard to something more appropriate. Beware that this may be a culture shift and that there is probably some history in the organization. Do your research to establish your starting point and work on your relationships and sales ability to gain acceptance for your proposed changes. Avoid making any value judgments in public.

◆ PERSONAL MORALS

An individual develops personal morals based on many factors—parents, religion, schools, peers, and even society itself (see Figure 10.1). You should be able to add to this list, considering those people and institutions that influenced your development. As you do, think of how all these influences help establish individual and organizational morals and ethics. Because there are so many variables, it is easy to understand why there are so many different points of view. The important thing to note is that everyone has a different background with respect to the development of their ethics. As a result, there often is no right or wrong answer to every situation. As with so many situations and issues, recognizing that ethics is an important component of the job is the first step toward preparation. You should be convinced to study and learn about ethics.

Ethics can be influenced by power, money, politics, and greed. Some people can justify their action based on some or all of these factors. Further, the pressure of these influences can cause otherwise normally ethical people to be blinded to the proper decision and do something they usually would not do.

◆ ETHICS AND YOUR JOB

As you ascend the ranks, many decisions are not so easy. Whereas a firefighter may enjoy clear right and wrong choices, higher-ranking officials find that issues are seldom black and white; almost all circumstances are gray. Situations are almost never exactly alike; there are always variables and extenuating circumstances and nuances to be considered.

You will be judged by the actions you take every day—on the simple routine activities and on the not-so-routine, even extremely complex, events. You need to be fair and impartial. Unfortunately, everyone's concept of fairness and impartiality is slightly different. Does this mean that fair and equal are the same? Consider the justice system. When a crime is committed and the responsible person is caught, there is a trial. A conviction means some punishment is to be given, often a term of imprisonment. Are all convicted individuals given the exact same sentence? Are judges given

FIGURE 10.1 ◆ Schools and religious institutions help individuals develop their ethics.

ranges (2 to 5 years for example) to consider in sentencing based on the individual circumstances of the crime? If so, is this fair? You should be able to make your own judgment regarding this question.

After you have made a "fair" judgment and taken corrective action, others will view your action based on their individual ethics and their perceptions. Further, you often do not have the luxury of disclosing all your information, because you may be bound by confidentiality rules. To further complicate the job of fire chief is the close-knit nature of the fire service. The bond among firefighters is strong, creating nearly unwavering support among the firefighters that may place them at odds with the fire chief, regardless of the facts. Often, when sides are taken, the chief can be left alone. Nevertheless, you must do what you think is right and live with your decisions. You will build your reputation each day with each decision, action, and choice. Conversely, your reputation can collapse like a house of cards with one act of indiscretion.

◆ DEPARTMENT PERSONNEL AND ETHICS

The second reason to learn and understand the basic guidelines of ethics is to help you manage, administrate, and better lead the organization. As mentioned earlier, people have different perspectives based on their backgrounds. If you don't have a general knowledge of these differences, you will not understand your employees and will not be able to fulfill your major role of leading the organization. The department members must make decisions and choices consistent with the ethics of the organizations, not their own beliefs. As mentioned earlier, different backgrounds, beliefs, and morals lead to different interpretations and decisions. People do not think the same, so it cannot be expected that their choices will be the same. Therefore, it is important to learn as much as possible about ethics and apply that knowledge to your role as the fire chief.

◆ SOURCES OF ETHICS

The development of ethics in individuals is influenced by many factors, including parents/family, schools, church/religion, culture, and peers. Other influences are specific to individuals depending on their own upbringing and influences. Your list may include military service, mass media, individual "heroes" (possibly sports stars, entertainers, or other celebrities), scouting, and the like.

If you examine the major contributors to the development of ethical beliefs and morals, you can see why the workforce is diverse and constantly changing. For example, once, the traditional nuclear family represented the majority of home life. Today there are many other types of "families." Single parents, grandparents raising children, gay couples adopting or having their own children are becoming more prevalent, so much so that the traditional family may not represent even 50 percent of homes.

Next, an examination of the current state of the school system reveals that schools are asked to address more than the basics presented years ago. Pressure on students to get good grades may erode integrity. Outside influences previously not experienced in the schools such as guns, violence, drugs, and sex now are part of the day-to-day experiences of many students. Additionally, the student population is growing more diverse, which affects peer pressure. Thus, you may begin to understand why the world is not the same as it used to be and why learning all you can becomes important in establishing the proper ethical climate in your organization. Based on educational background alone, the people you are hiring today do not think like you!

Each person has an individual set of beliefs and values. **Beliefs** are statements, principles, or doctrines that the person accepts as true. The **values** are the accepted principles or standards of the individual or group. From the discussion so far you realize that everyone does not think alike, and each organization has its own culture, set of beliefs, and values. Problems occur when individual ethics and organizational ethics don't match. These ethical "potholes" are like potholes in the road—they need to be fixed when identified and avoided when encountered. As much as you may try to avoid them, sometimes you will hit the "pothole." When an ethical issue is presented it must be dealt with quickly and in an ethical manner.

beliefs
Statements, principles, or doctrines that a person accepts as true

values
The accepted principles or standards of an individual or group

All in all, most people want to do the right thing. Sometimes they just need to know what the right thing is. Rarely, especially in the fire service, do you find someone blatantly going against acceptable practices. (If you have been around for awhile, you know that there are individuals who just "don't get it.") If you hire the right people and give them the tools to do the job, nearly all will make every effort to do the right thing. Yet, as has been discussed, ethics are complex and ever changing (think of the issues that computers have raised). Ethical solutions cannot always be recognized until it is too late.

Surely you have recognized that the workforce has changed, particularly in the fire service. What was once a predominantly white male domain now includes more females and other races, religions, and cultures. It is easy to understand how issues in a homogeneous setting can be understood by all, because they have the same foundation. Likewise, you can understand how things can be different, sometimes significantly and in other circumstances more subtly. The diversity of backgrounds affects decision-making ability. Consider the differences of opinion between people in their twenties and those in their forties or fifties. You cannot expect people with these differences to make similar choices or decisions. This is not to imply that one group is right and another wrong. It just means that as a result of differences, people may see things differently. Hence, the issue of ethics looms. As a side note, you should not take a negative view of the younger generation because they think differently. You most likely think differently than the generation before you, or at least you did when you were younger! As you age, you change responsibilities, and life-changing experiences can alter you viewpoints.

◆ APPLICATIONS

Think of some examples of ethical situations and the potential for problems:

- An elected official wants a favor.
- Vendors send gifts to the department.
- Funds are spent on questionable items.
- Gifts are received from the public.
- While shopping for lunch, firefighters do personal shopping.

The list can go on. In each case there are degrees of seriousness and mitigating circumstances. Take for example the situation of vendors' sending gifts. Does it matter if the value is low? Can someone accept a pen, baseball hat, or other small item?

Is there a difference if it is given at a trade show or directly at the fire department? If so, where do you draw the line? At what point is the gift considered inappropriate? Is it based on dollar value? Is it reasonable (or even possible) to have a zero-tolerance policy? This is a very simple example, and even it can generate a list of questions. Again, individuals may answer the questions differently, and depending on the established rules, the answers may be acceptable. Because there are different possibilities, these are questions that need to be answered before becoming an issue. If not, they can affect you, other members, or the entire organization.

◆ PERSONAL PREPARATION

What can you do? This is a significant issue that needs your attention. Acknowledgement that ethics are important is an appropriate first step. Educate yourself. There are many sources of information. The International Association of Fire Chiefs, International City/County Managers Association, and National League of Cities all have information regarding ethics and ethics policies. There are also other resources such as the American Society for Public Administration or the Markkula Center for Applied Ethics at Santa Clara University. You can check out these and others on the Internet for more information.

Find out what you can do to both protect yourself and understand the issues that you will face in your department. If there is a written policy for your community or organization, read it. If there isn't one, create it. There are many examples to provide you guidance. Also, do not be afraid to broach the subject with your boss. Let him or her know that you are aware of the potential for problems when ethical issues present themselves, and you want to act in accordance with the organizations values.

You will constantly be challenged by ethical situations, sometimes on short notice, sometimes long term. Be aware that situations where self-interest is at stake tend to dull one's ethical sensitivities. Also be aware of the influence of others, as it is easy to be lulled into certain actions based on the success of others. Ethical behavior may make it difficult to compete with those stretching the boundaries of acceptable practices. Do not change your beliefs regardless of what is happening around you.

Individuals can always use help in developing their skills, knowledge, and abilities. With respect to established ethics for personal behavior, you certainly have a good foundation that has probably served you well. You can talk to others to get different perspectives. Professional associations can also be a valuable resource. For example, the International Association of Fire Chiefs has a model code of ethics for fire chiefs (see Figure 10.2). This is just another option for you to consider.

You can do a few things for your individual development and protection, such as the following:

- Know the rules of your organization and society.
- Know what is expected for situations whether or not policies exist.
- Practice making good ethical decisions; play what-if games.
- Sweat the small stuff.
- Seek the opinions of others.
- Do not take advantage of loopholes.

International Association of Fire Chiefs

Fire Chiefs' Code of Ethics

The purpose of the International Association of Fire Chiefs is to actively support the advancement of the Fire Service which is dedicated solely to the protection and preservation of life and property against fire and other emergencies coming under the jurisdiction of the Fire Service. Towards this endeavor, every member of the International Association of Fire Chiefs shall, with due deliberation, live according to ethical principles consistent with professional conduct and shall:

Maintain the highest standards of personal integrity, be honest and straightforward in dealings with others, and avoid conflicts of interest.

Place the public's safety and welfare and the safety of firefighters above all other concerns; be supportive of training and education which promote safer living and occupational conduct and habits

Ensure that the lifesaving services offered under the member's direction be provided fairly and equitably to all without regard to other considerations.

Be mindful of the needs of peers and subordinates and assist them freely in developing their skills, abilities, and talents to the fullest extent; offer encouragement to those trying to better themselves and the Fire Service.

Foster creativity and be open to consider innovations that may better enable the performance of our duties and responsibilities.

FIGURE 10.2

- Don't use your position for personal gain.
- Be conscious of the appearance of impropriety.
- Obey all laws.

◆ ORGANIZATIONAL POLICIES

Do you have a published ethics policy for your organization? This is a formal statement of your organization's values and is the document that should govern the actions of your entire membership. Include in it guidelines for decision making. The policy should include the entire governmental structure, not just the fire department. If the government agency as a whole does not have or desire a policy, a simple one should be developed. There are many great examples to review so that you can develop your own. Once you have done this, it is always a good idea to have it reviewed by an attorney (one who may be aware of state and federal laws as well as local ordinances). See Figure 10.3 for a sample of what a code of ethics should include.

◆ ETHICAL DECISION MAKING

Regardless of your circumstances, you are likely to be faced with an ethical dilemma during your career. Most likely your instincts will lead you to the right decision. On occasion, however, you may come on a more complicated case. For these instances, you need a system to help you make the correct decision. Discuss the situation with your

CODE OF ETHICS

CITY OF ABC

Section 1. Public Policy

It is hereby declared to be the policy of the City that all officials and employees must commit themselves to avoid conflicts between their private interests and those of the general public whom they serve. To enhance the faith of the people and the integrity and impartiality of all officials and employees of the City, it is necessary that adequate guidelines be provided for separating their roles as private citizens from their roles as public servants. Where government is based on the consent of the governed, every citizen is entitled to have complete confidence in the integrity of his government. Each individual official, employee, or advisor of government must help to earn, and must honor, that trust by his own integrity and conduct in all official duties and actions.

Section 2. Definitions; as used in this Code

I. "City official/employee" means a person elected, appointed, or otherwise serving in any capacity with the City or in any position which is established by the City Charter or by City ordinance which involves the exercise of a public power, trust, or duty. The term includes any official or employee of the City who receives compensation on a permanent, regularly scheduled, continuing basis from the City, including persons who serve on advisory boards and commissions.

II. "Decision making" means exercising public power to adopt laws, regulations, or standards, render quasi-judicial decisions, establish executive policy, or determine questions involving substantial discretion.

III. "Substantial" means anything of significant worth and importance, or of considerable value, as distinguished from something with little value, social tokenism, or merely nominal.

IV. "Compensation" means any money, property, thing of value or benefit conferred upon or received by any person in return for services rendered or to be rendered to himself or another.

V. "Official duties" or "Official action" means a decision, recommendation, approval, disapproval, or other action or failure to act which involves the use of discretionary authority.

Section 3. A code of Conflict of Interest and Ethical Conduct is hereby promulgated as follows:

Gratuities

I. No City official/employee of the City shall solicit, accept, or receive, directly or indirectly, any substantial gift, whether in the form of money, service, loan, travel, entertainment, hospitality, thing or promise, or in any other form, under circumstances in which it can reasonably be inferred that the gift is intended to influence him or her in the performance of their official duties or is intended as a reward for any official action on their part.

Preferential Treatment

II. No City official/employee of the City shall use, or attempt to use, their official position to unreasonably secure, request, or grant, any privileges, exemptions, advantages, contracts, or preferential treatment for themselves or others.

Use of Information

III. No City official/employee of the City who acquires information in the course of their official duties, which information by law or policy is not available at the time to the general public, shall use such information to further the private economic interests of themselves or anyone else.

Full Disclosure

IV. No City official/employee of the City shall participate, as an agent or representative of a City agency, in approving, disapproving, voting, abstaining from voting, recommending, or otherwise acting upon any matter in which his or her has a direct financial interest without disclosing the full nature and extent of his or her interest. Such a disclosure should be made before the time to perform his or her duty or concurrently with that performance. If the officer or employee is a member of a decision-making or advising body, he or she should make disclosure to the chairperson and other members of the body on the official record. Otherwise, a disclosure would be appropriately addressed by an appointed officer or employee to the supervisory head of their organization, or by an elected officer to the general public.

Outside Business Dealings

V. No City official/employee of the City shall engage in or accept employment or render services for a private or public interest when that employment or service is incompatible or in conflict with the discharge of the official or employee's official duties or when that employment may tend to impair his or her independence of judgment or action in the performance of official duties.

VI. No City official/employee shall engage in a business transaction in which the public, City official, or employee may profit from his or her official position or authority or benefit financially from confidential information which the public official or employee has obtained or may obtain by reason of that position or authority.

Doing Business with the City

VII. No City official/employee shall engage in business with the City, directly or indirectly, without filing a complete disclosure statement for each business activity and on an annual basis, in accordance with the City Charter.

Suppression of Public Information

VIII. No City official/employee of the City shall suppress any public City report, document, or other information available to the general public because it might tend to unfavorably affect his or her private financial interest.

Use of City Property

IX. No City official/employee of the City shall directly or indirectly, make use of or permit others to make use of City property of any kind for purely personal gain. City officials/employees should protect and conserve all City property including equipment and supplies entrusted or issued to them.

Section 4. Intention of Code

It is the intention from Section 3 above that City officials and employees avoid any action, whether or not specifically prohibited by Section 3, which might result in, or create the appearance of:

I. Using public employment for private gain;
II. Giving preferential treatment to any organization or person;
III. Impeding City efficiency or economy;
IV. Losing complete independence or impartiality of action;
V. Making a City decision outside official channels;
VI. Affecting adversely the confidence of the public or integrity of the City government; or
VII. Accepting preferential treatment in use of City property

The Conflict of Interest and Ethical Code is intended to be preventive in nature rather than punitive. It should not be construed to interfere or abrogate in any way the provisions of any State Statutes, the City Charter, and/or Ordinances.

(*continued*)

This declaration of policy is also not intended to apply to contributions to political campaigns or to prevent any official/employee of the City from receiving compensation for work performed or services rendered not involving City business and purely on his or her own time as a private citizen.

Section 5. Violation, Enforcement, and Advisory Opinions

I. All matters concerning the Conflict of Interest and Ethical Code shall be directed to one of the two following controlling authorities depending upon the employment status of the City official/employee involved, or group concerned, and the nature of the action requested:

   a. Elected and appointed officials of the City to the Mayor, City Council, and City Attorney
   b. Appointed employees, full and part-time, of the City to the City Manager and City Attorney

II. The above listed authorities, when requested, shall take appropriate action upon any complaint, request for information, or otherwise resolve matters concerning Conflict of Interest and the Ethical Code policy of the City. The appropriate action to be taken in any individual case shall be at the discretion of the controlling authority involved, which may include but is not limited to any of the following:

   a. Referral of the matter to a higher authority
   b. Pursuing further investigation by the controlling authority
   c. Taking appropriate disciplinary action in accordance with the City Code, State Law, or the regulations or policy of any City Department
   d. Deeming no action to be required
   e. Pursuing such other course of action which is reasonable, just, and appropriate under the circumstances

III. The above listed controlling authorities may render written advisory opinions, when deemed appropriate, interpreting the Conflict of Interest and Ethical Code of Conduct as set forth in Section 3 above. Any City official/employee may seek guidance from the controlling authority upon written request on questions directly relating to the propriety of his or her conduct as officials and employees. Each written request and advisory opinion shall be confidential unless released by the requester.

   a. Request for opinions shall be in writing
   b. Advisory opinions may include guidance to any employee on questions as to:

      1. Whether an identifiable conflict exists between his/her personal interests or obligations and his/her official duties
      2. Whether his/her participation in his/her official capacity would involve discretionary judgment with significant affect on the disposition of the matter in conflict
      3. What degree his/her personal interest exceeds that of other persons who belong to the same economic group or general class
      4. Whether the result of the potential conflict is substantial or constitutes a real threat to the independence of his/her judgment
      5. Whether he/she possesses certain knowledge or know-how which the City agency he/she serves will require to achieve a sound decision
      6. What effect his/her participation under the circumstances would have on the confidence of the people in the impartiality of his/her City officials and employees
      7. Whether a disclosure of his/her personal interests would be advisable, and, if so, how such disclosure should be made so as to safeguard the public interest
      8. Whether it would operate in the best interest of the people for him/her to withdraw or abstain from participation or to direct or pursue a particular course of action in the matter.

FIGURE 10.3

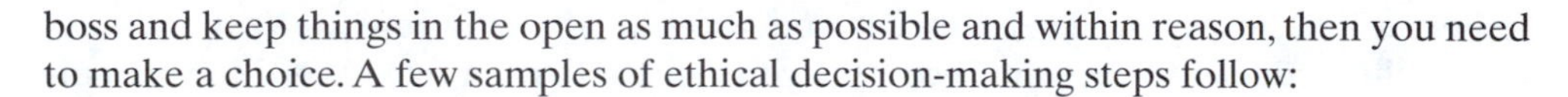
boss and keep things in the open as much as possible and within reason, then you need to make a choice. A few samples of ethical decision-making steps follow:

Option 1

- What are the facts?
- What are the options?
- What ethical values and principles are at stake?
- Who will be affected by your decision?
- Who needs to help determine the course of action?
- Which options benefit which groups?
- Are fundamental rights involved?
- Are your personal values in conflict with the organization's values?
- Are they in conflict with those of others who must participate in the decision or with those of the affected groups?
- What are the short- and long-term costs of your decision?
- What adjustments do you need to make based on unexpected consequences or new information?

Option 2

- Recognize the ethical issue.
  - Is the issue personal, interpersonal, institutional or societal?
  - Why is it an ethical issue?
- Start the decision process.
  - Gather facts. Who is involved?
  - What choices are available? What would others think of your choices (especially people you respect)?
- Evaluate the alternatives.
  - Which has the best outcome?
  - Which choice is within your moral beliefs?
  - What is best for all individuals involved?
  - Which choice is fairest?
  - Which choice promotes common good?
  - Which choice would establish the best precedent?
- Make the decision.
  - Consider the various points of view.
- Evaluate the outcome.
  - Did the action turn out as expected or produce the desired result?
  - If not, can adjustments be made?

◆ DIFFERING VIEWS ON ISSUES

As discussed earlier, ethics vary because of the diversity of ages, genders, races, cultures, and individual backgrounds. As organizations diversify, ethics, values, beliefs, and judgments become mixed. Whereas most can agree on the major ethical issues, it is often the lesser ones that present the most challenges. As an example, consider the use of sick leave. Many believe that you use sick leave as needed. Others believe sick leave is earned and available for taking whenever desired. Some employees bank whatever they can, go to work every day, and do not use the leave unless they are really sick.

Others will use every available sick day and have few, if any, in reserve. Without making a value judgment as to who is correct, you can see that there will be different perspectives. You also can probably identify many more examples from your experience that further demonstrate the differences in beliefs.

Even with their differences, people generally want to do the right thing. As chief, you need to do what you can to help them make the right choices by doing the following:

- Publish a policy.
- Train/educate formally and informally.
- Hold discussions on ethical issues.
- Set a good example.

Create the proper climate for ethical behavior. Evaluate your organizational culture—where it is and where you want it to be.

Train and educate your personnel with respect to the policy. You may need outside help in delivering this information, because it is sometimes difficult to discuss ethics close to home. Presenting general information may raise unnecessary questions. Although there are likely to be differing opinions, you still need to set the tone and communicate your requirements. Increase employees' awareness of and encourage open discussions about ethical issues, both simple and more complex. Make sure that everyone in the organization knows they will be held accountable for following the policy.

◆ TACTICS

Ethical decision making can be very taxing. There are a few things that you can do when faced with an ethical situation requiring a choice: Deal with the issue in the open, unless it involves obvious confidential and classified information. This helps clear the air and prevent accusations of impropriety or hidden agendas. In certain circumstances, openness can instantly clear the air. For example, should you be approached to provide a "favor" that stretches the law or policy, simply repeat the request to the person. Most likely an individual does not want to make an illegal request, or you may have misunderstood the issue. Openness clarifies the situation.

Follow the decision-making steps outlined previously, and consult others who may offer a different perspective. People can always use help and guidance. For example, Rotary International, a service group, has what it calls a "Four-Way Test" that it uses to end its meetings. It is simple and can be used to guide ethical decision making (see Figure 10.4). Another approach is to consider a couple simple questions: Would you want to see your decision, action, or inaction on the evening news? Would you want your mother to see your decision, actions, or inactions on the evening news? These are just a couple suggestions; you have some thoughts of your own.

Rotary Four-Way Test

- Is it the truth?
- Is it fair to all concerned?
- Will it build goodwill and better friendships?
- Will it be beneficial to all concerned?

FIGURE 10.4

◆ SUMMARY

Ethics and morals play an important role in organizations. Some issues are simple, and others can be extremely complex. There are many gray areas that allow you to choose from among different options in making decisions. Prepare yourself and your department. This can help reduce exposure to liability. Adopt a formal policy. Know the simplest ways to deal with ethical choices: Is it legal? Is it balanced? and How do you feel afterward? Do the right things, for the right reasons, even when no one is looking.

Activities

1. Review your department/city code of ethics. If you don't have one, develop one. If you have one, would you change anything? Have all your members received training on the document? Develop a plan to ensure that all employees understand the intent of the policy.
2. You receive an anonymous call at your office in which the caller states that he saw one of your fire inspectors eating at a local restaurant. The caller tells you that a check for the meal was never presented, and your officer did not pay. What will you do? Would it matter whether the inspector was on duty?
3. A restaurant in your community is known for giving police and firefighters a 50 percent discount. It has been brought to your attention. What will you do?
4. Does your community have any policies regarding emerging technology? Is there a policy on cell phone usage? How about on computer usage, specifically e-mail and the Internet? If there is no policy, discuss the potential issues that may arise. If there is, evaluate its currency based on modern technology.
5. A local politician approaches you regarding a fire safety issue at a restaurant. A fire department inspector has cited the restaurant for fire code violations. The owner is a friend of the politician and also employs two children of the politician. The owner would like some leeway on the violations. What will you do?

◆ CASE STUDY

WINE DELIVERED AFTER FIRE

Your department responded to a fire in a strip mall. The fire department units did a great job and made a significant save that preserved most of the businesses. One of the businesses was a convenience store with a significant stock of fine wines. The next evening, the owner of the store stops by the closest fire station and leaves three cases of fine wine at the door with a personal note thanking the department for the excellent work. When you arrive at work the next morning, you are told of this by the officer in charge (whose shift wasn't even at the fire), and the wine is stacked in your office.

What will you do?

CHAPTER 11

Technology

KEY TERMS

technology, p. 132

There is no doubt that technology has changed the way fire departments are managed and administered, but the obvious benefits of technological advances do not necessarily make the job of fire chief any easier. **Technology**, which is the study, development, and application of devices, machines, and techniques, makes things faster and more efficient in many cases. Embracing technology can ultimately lead to better productivity; but getting to the point at which the processes are beneficial may not be as easy as it may appear.

technology
The study, development, and application of devices, machines, and techniques

It is obvious that technology has advanced dramatically in the last 10 to 15 years and that changes are occurring at an accelerating pace. Technology has influenced everyday life and certainly the day-to-day operations of the fire department. It is absolutely essential to embrace technology and use it to its fullest advantage in operating your organization, even though the rapid changes may make planning more difficult or challenging. Your organization may need to become more responsive and adaptable, since plans can become outdated before they are implemented.

◆ TECHNOLOGICAL ADVANCES

Technology has changed the way the fire service does business on the scene of an emergency and also in the way staff functions including management, administration, fire prevention, and training are performed. The number and speed of the changes make it difficult for any individual to keep up. If you are the fire chief with all the responsibilities listed in the previous chapters, how can you possibly know and understand everything?

Consider the technological advancements that have been made to fire departments since you joined, in areas such as apparatus, equipment, reporting (i.e., computers), and communications. All are much different than they once were, and certainly will be different in the coming years. Technology has enabled tasks to be done faster and, in most instances, better. As fire chief you must understand a little bit about technology to be able to add to the resources needed to provide the service and also to determine which changes may not significantly improve the level of service.

You are challenged, also, to deal with people who know a lot more about technology than you do. In many municipalities, there is at least one information technology professional, and there may even be an IT department. Further, some of the firefighters in your organization may also be quite knowledgeable in some aspects of modern technology. Sometimes, these "unofficial" experts know enough to push the envelope with respect to utilizing technology.

The good news is that you do not need to fully understand all the technology heading your way. Do you know how your car works? Unless you are an engineer or mechanic—and even those credentials may not be enough to understand modern automobiles—you probably don't know the inner workings of your vehicle. You get in the vehicle, start it up, drive away, and trust that those who know how things work have done their job and you will get where you need to go. There are many elements of the fire service that may be beyond your understanding as well. It doesn't matter. You need some basic understanding, the ability to do research when needed, and personnel whom you trust who can do the work.

◆ UNDERSTANDING TECHNOLOGY

What does a chief need to know? To answer this question it may be instructive to look at the operation of an automobile as an example. A person is able to operate a vehicle with very minimal knowledge of the inner workings of the engine and other components that make the vehicle go. The driver needs to know how to start and stop the vehicle, maneuver it as it moves, and operate the essential components within the vehicle. There are many parts of the vehicle of which most drivers are unaware, or of how they function. Regardless, the driver can use the vehicle for transport to the desired destination. There are many models and makes of vehicles, each offering advantages and disadvantages within different price ranges. Technology is similar to that. The fire chief does not need to know the inner workings of a computer but needs to know its capabilities and the level of service needed within the department.

Depending on the amount of time you have spent in the fire service, you can develop quite a list of changes that have occurred owing to technological advances. You also can probably recall the challenges that were created when the technology was introduced. Not everyone embraces the new way of doing things, nor are all new things better than the old way of doing something. You have two challenges: determining whether the technology will benefit the organization and how to implement the new technology.

Technology can offer new services, enhance existing services, or support some of the basic things done in the fire department. A few years ago, carbon monoxide alarms entered the mass market. This created new challenges in the short term for the fire department. What did you do when someone called 911 to report an alarm? You had to respond. Did you have the right tools to assess the emergency? Do you remember with whom you responded? Were they adequately trained? You can see where this line of questioning is leading. As new things are created, new challenges are presented. Carbon monoxide detectors created more responses. These responses led to contact with the taxpayers. To be most effective, fire departments needed to learn what to do when they received these calls. Fire departments acquired better monitors, trained their personnel, established or changed their procedures, and looked for ways to improve service to the community based on the newer technology.

◆ COMPUTERS

The use of computers for record keeping allows fire departments to collect more data and retrieve it faster. Computers also facilitate better analysis of the service levels being provided and make it possible to plan for the future with more information available. Stored data such as preemergency plans and other documentation can also be retrieved while en route to an emergency, and incident action plans can be developed during an incident (see Figure 11.1).

What started with one computer with some basic entry of information has evolved into complicated networks with loads of information being entered every day. Fire chiefs must evaluate hardware and software; they must educate their personnel; they must have an idea how to use the information they are gathering, and they also need to understand that all this information is subject to disclosure during legal action and Freedom of Information Act (FOIA) requests. Technology may save time in some ways but create other work in the process.

FIGURE 11.1 ◆ Mobile Data Computers get information to responders that improves the operation.

EMERGENCY SCENE ADVANCES

Think of the things that have changed that affect your response to emergencies. Many things that were introduced that were designed to improve response and performance, yet in most cases, the new way was not embraced. When the horses were replaced, it took time to gain acceptance of the new apparatus. When self-contained breathing apparatus was introduced, the air packs were only used on the "bad" fires. It took many years to mandate use on day-to-day fires, with veteran firefighters the most reluctant to use SCBA. Looking back, it is sometimes hard to believe that there was resistance. Once the modern technology is accepted, it is hard to understand why there was such opposition to the change.

Relatively recently, thermal imaging cameras were introduced as an additional tool in the firefighting arsenal. Fire chiefs and their staff are expected to develop policies, procedures, and training programs to maximize their benefit. When these cameras were initially introduced, they were marketed as an aid to search and rescue. They could "see" through smoke, making it easier to locate victims and expedite rescues. Since then firefighters have found many more uses such as finding hot spots, locating hidden fires, and even determining the level of a hazardous material in a container. As the thermal imagers became more commonplace, the comfort level of the fire fighters increased and they found more uses.

The challenges to the fire chief are to be aware of the new technology, understand its use, and determine if the benefit justifies the cost (see Figure 11.2). Once the decision is made to expand the use of technology and make the purchase, there needs to be a system to introduce it to the members of the department. Returning to the example of the thermal imaging camera, when it was first introduced, the cost was significant for most departments. Even after many departments determined that there would be a benefit, finding the resources, that is, money, to make the purchase was difficult if not impossible. A combination of private donations, public pressure, political support, and a hard sell by the fire chief got this valuable tool into the hands of the firefighters. As more departments embraced the device, more were built and the cost came down, following the law of supply and demand. Soon the cameras became commonplace, which led to more policies and procedures. Now it is almost expected that all departments have at least one and use it regularly.

The following is a partial list of the newer technological advances that have changed, to some extent, the way firefighters train and attack fire:

- Computers
- Fire apparatus
- Computer-aided dispatch
- Newer SCBA
- Newer turnout gear
- Accountability systems
- EMS equipment
- Thermal imaging cameras
- Ventilation fans
- Power tools
- Extrication tools
- Technical rescue equipment
- Compressed air foam

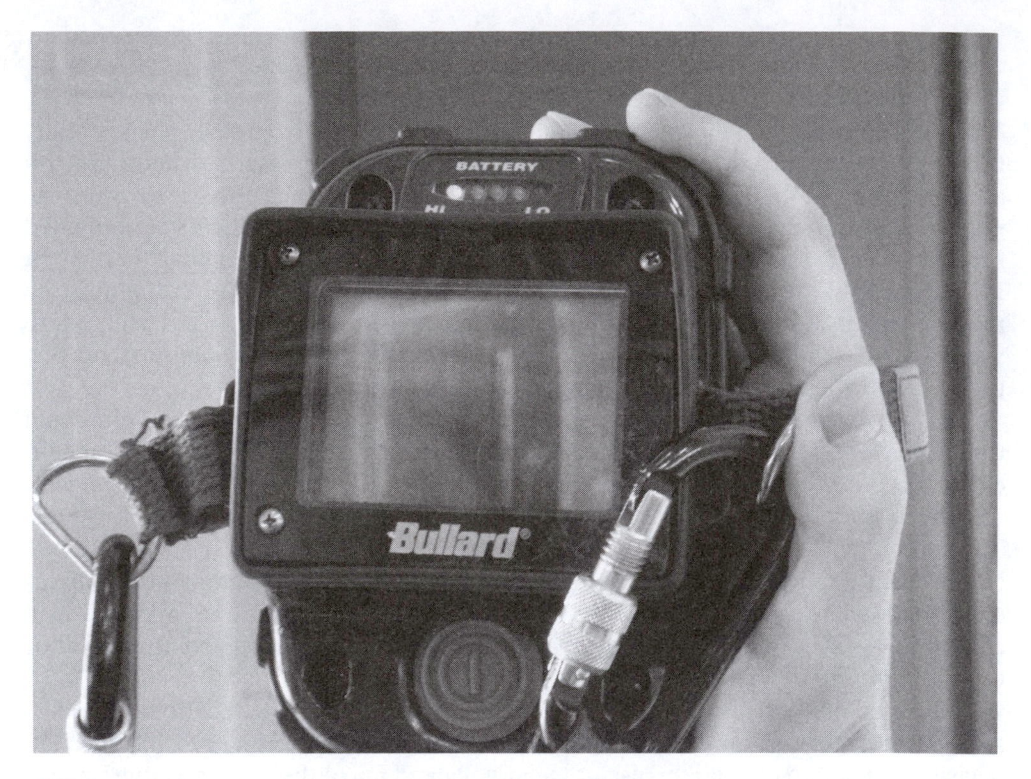

FIGURE 11.2 ◆ Technology on the scene aids in tactics and safety.

As mentioned previously, even items that have been around seemingly forever are changing rapidly, for example, fire trucks. Although their basic premise remains the same—get water on the base of the fire—changes in equipment affect the firefighting approach.

◆ KEEPING UP

It is virtually impossible for one person to stay current, yet if that one person is you, the fire chief, you will be expected to have enough knowledge to make an intelligent decision to acquire these technological advances. You will then need to find the money to purchase it and convince your personnel to embrace the new "stuff" and figure out the best way to use it. Finally, you will be judged on its success. That sounds easy!

New technology that promises to save lives and property places tremendous pressure on the fire chief. If the technology has been around long enough to be considered mainstream, there is an expectation that every department will have one. If disaster strikes, in the form of a fatality or serious fire, and the department did not have what many would consider "basic" equipment, the chief and department could be subject to extreme criticism for failure to provide the basics. Often, the emotions surrounding the incident create opportunities to question the abilities of the fire department and its leader. Had the technology been used, the outcome might have been different. Even though it is difficult to predict an outcome based on the lack of equipment, many will think that everything possible was not done. Such a situation opens the door to people looking for a reason to criticize the department or to those seeking an explanation as to why something bad happened.

◆ QUESTIONS TO ASK

In other cases, new technology is so new that it is not proven in the industry. This, too, presents challenges. Emerging industries want to sell their products. Some will turn out to be excellent and will have a place in the emergency service. Some will end up on the auction block. The questions to ask yourself are: What is the value? What will work? What is the cost? Is it wise to wait? What proof will you need that the product is beneficial? Can you convince your employees of the value? At some point, technology may even affect staffing levels. How will you address this with your labor group and your policy makers?

Fire chiefs are made aware of new technology through a variety of sources, through other department members, through vendors trying to "educate" you, through your network, while attending trade shows, reading professional journals, or on the Internet (which in itself could be considered relatively new technology!). Often, due to the special nature of emergency equipment, there are few experts within your inner circle, and your personnel may not have the background to evaluate the benefits. Normally, information technology (IT) experts are on top of the changes in the IT field; however, the IT experts probably do not understand the emergency services or your operation well enough to make a qualified recommendation. This is why a strong partnership between you (and the department) and the IT specialists is so important. Each has knowledge to offer, and only through a good working relationship will you get the equipment that works for your department. If your organization is to be one of the first with newer technology, there will be some risks to your reputation. If it works out, you are the hero. If not, you have wasted limited resources.

You should have a system in place to evaluate technology, including a cost/benefit analysis for considering the cost of the item and its potential impact on service delivery. Will the product deliver what it promises, and will the enhancement be sufficient to justify the cost? This analysis is not always easy, especially if you are on the leading edge of change. Take for example the use of computers during emergency response. There is no doubt that they can provide an endless amount of information, but it comes at a cost. Although the computer can access the information, someone needs to program the computer. Obviously there is a cost to this. On the user end, you must ask whether the information being provided is worth this cost in time and money. If your community and its street system are relatively simple and logical, is the cost of providing that information low enough based on the marginal improvement in information that can be provided? That is ultimately for you to decide. The technology is available, but the improvement to service may not justify the cost.

◆ FINDING OUT WHAT IS NEW

First and foremost you need to know what is available and likely to be available. You do this through all your contact methods. If you are a large enough organization, your staff should be bringing you information and ideas. If this is the case, you need to create a culture that encourages new ideas. Your members need to know it is okay to suggest new things, but this is not to say that you blindly accept all recommendations. Your members need to know what you expect regarding evidence that the new product will work. Your members need to know that you will ask questions, and that these are not to be interpreted as a challenge to the person bringing the idea but part of your responsibility in overseeing the department.

There is tremendous value in attending trade shows and reading professional journals to learn about new technology. For some of the newer products, you may need to expand your search outside the normal fire service resources. Vendors should also be considered a resource with respect to emerging technology. Although vendors have an obvious bias toward the product they are selling, they also have more knowledge than you do. A couple of suggestions are to build relationships with them so you gain some level of trust, and utilize multiple vendors to compare information.

◆ WORKING WITH OTHERS

Because of the nature of technology, there is probably no other part of the job of fire chief that requires more trust in others to make decisions. This applies not only to equipment that helps on the scene of emergencies but also to that used in the day-to-day administration of the department. Departments rely on computers to do so much, and they are capable of doing almost everything. Owing to the other requirements of the job, few fire chiefs can keep up with computer technology. Further, depending on the size and complexity of your community, you may be reliant on others in your local government to provide the IT solutions.

Departments today cannot operate without telephones, e-mail, and the computer services that provide for human resource management, purchasing, budgeting, and payroll, to name a few tasks. Often, the fire chief does not get a vote as to what computers and software are used for the normal activities—those that are considered business functions, not emergency service needs. That means your phone system, computer network, e-mail system, and business software are probably recommended or provided without your input. This is not necessarily bad, because you don't know the details anyway, but they do have an impact on your services and administrative responsibilities.

For example, a financial software package may be provided for you. You will be expected to use the system for anything related to financial issues including budgeting, purchasing, payroll, and maybe a few other services. Although you probably did not have any significant input on what software is used, it has the potential to change the way you have done things in the past. You will be expected not only to use the software but to adjust some of your administrative practices to "fit" the new programs.

Computer-aided dispatch has changed the way many dispatch centers function. Modern systems allow dispatchers to take the call, enter what they know into a computer, and based on the address and caller-provided information, dispatch the appropriate units and provide prearrival instructions. Dispatchers can then provide much more information to the responding crews. Of course, this system has changed the way dispatch centers operate. It requires more training and may slow processing time, as some dispatchers become reliant on the system to tell them what to do. They need to enter the information before they can act. On the fire side, fire personnel, too, may become reliant on the computer. This is fine as long as all is working. When there is a glitch, the old way may be needed. Again, training is needed and there must be a backup.

◆ OBTAINING SUPPORT AND FUNDING

Assuming that others control your resources and your job is to influence spending decisions, one of the challenges you are likely to face is getting the support to spend money on new technology. If you are on the leading edge and one of the first to try something, expect questions and concerns to be raised. If the equipment is emergency specific, those outside your department will know very little; they will rely on your expertise. But first, you will need to explain the benefits with respect to the cost so that laypersons will understand how the product will work. If the item is more commonplace, such as a computer, expect everyone to be an expert, and expect more questions than you will get for specific specialty items.

◆ POLICIES AND PROCEDURES

With new technology will come new challenges to policies and procedures and the potential for abuse. Until recently, you would not have imagined that a cell phone use policy was needed. Now, with almost everyone owning a phone there is a potential that your members' cell phone use will create operational problems. Cell phones also can take videos and still pictures. Because the technology exists, you can expect someone to use it. Therefore, you will probably need a policy that limits its use.

The potential for abuse also exists with e-mail and Internet use. You will also need a policy to regulate these, as inappropriate e-mails and Internet use can cause problems and embarrassment. Your members need to know what is expected and the consequences of violations. In addition, you need to be aware of your limitations. You must not abuse the system or you will find yourself in hot water!

Regardless of what technology you are using to enhance your service, consider the need to establish control through policies and procedures. You can draft your own, borrow ideas from other fire departments, or consider other governmental agencies. Because things change so rapidly, you may not be able to find policy samples that address your concerns within the fire service, so consider sources in the private sector. Communicate the policy to your personnel in as many ways as possible, and let them know the consequences of inappropriate behavior.

◆ RELATIONSHIPS

As in so many other areas, relationships are critical. You cannot know all that there is to know about technology, so you are reliant on others. You need to build strong relationships with those responsible for technology in your organization, both inside and out the organization. A level of competence is necessary in your department to help with internal matters, but IT specialists outside the department but within the governmental structure most likely are much more technologically competent than you or your people. You will be reliant on their expertise, so you need to get to know them. They control much of what you do and may be the only ones who can fix systems when they don't work as they should. Do whatever you can to build up a trust that brings them in as partners so they are looking out for the best interests of the department. You will then be able to request support from them on technological issues and advancements that you know are good for your profession.

◆ TECHNOLOGY OUTSIDE YOUR ORGANIZATION

Not all technological advances affect your organization from the inside. Sometimes, external events will affect your department and the way you conduct your business. Even though you may not have a particular technology, someone else may. There may be an expectation that your organization could be using more technology to become more efficient and effective. You need to be prepared to respond should someone ask why you are not using all the tools available. Knowledge of the technical world can help you look for better ways to provide service, and the increased awareness can keep you out of trouble. For example, cameras are everywhere. You and your personnel can be captured on video or stills at any time, and those images can be sent anywhere instantly (see Figure 11.3). You and your personnel need to remember this and perform accordingly.

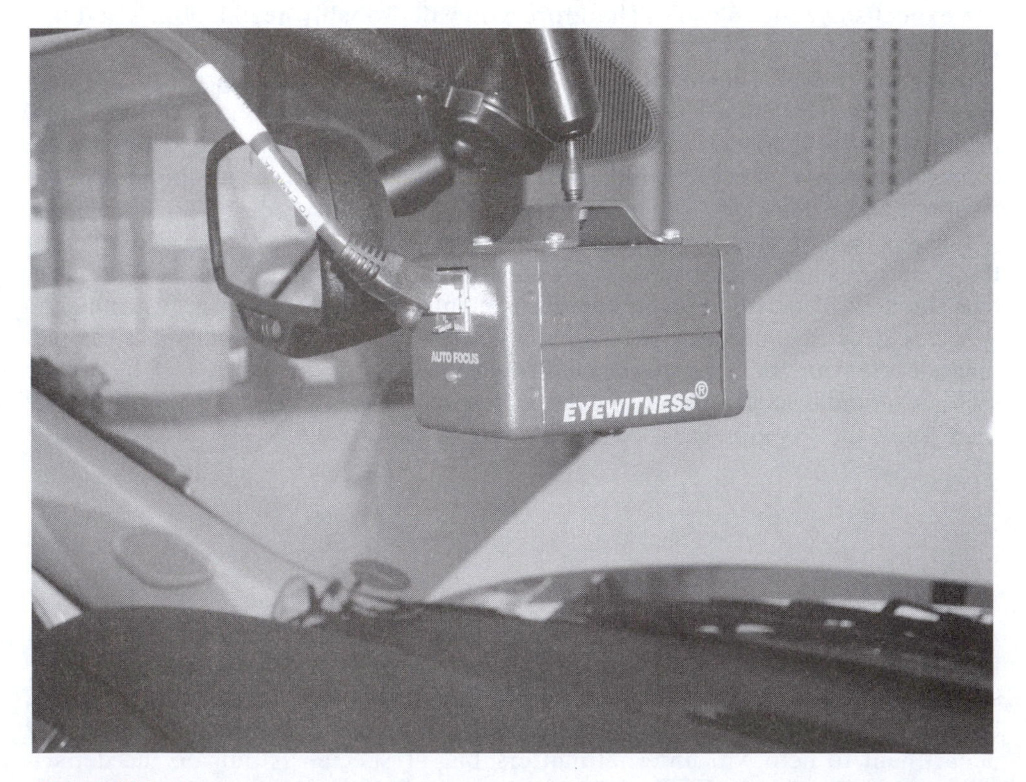

FIGURE 11.3 ◆ Cameras can be everywhere—just another reminder to do your best, because someone may be watching.

You may or may not know a great deal about technology. That does not matter. The world has become reliant on technology in general and computers in particular and all the things related to them. You need to learn the basics so you have a general understanding and establish relationships with those who are more competent. Understand that the challenges are similar to most of your other responsibilities as fire chief—you need good people to help you. You are then responsible to make the right choices based on all the information available to you.

ACTIVITIES

1. Identify the IT people who contribute to your organization and their responsibilities with emerging technology.
2. Develop a system for evaluating new technology, that is, a cost/benefit analysis.
3. Identify changes in technology in your department in the last 5 years. Discuss how the changes occurred and how they were accepted.
4. Your department is pursuing technology that may affect staffing levels. Develop a strategy for the new idea to gain acceptance in your organization.
5. Your IT department has implemented new software that will change how you administer your budget. What will you do?

◆ CASE STUDY

CHANGING A FIRE RECORDS MANAGEMENT SYSTEM

You are not happy with your records management system. Through your research you find a much better product that would be much more beneficial to your organization. The current product is outdated and does not provide the information you require. Entry is also labor intensive and the software is not user friendly. The IT department of your community has final approval authority of all computer systems used in the city and is typically slow to approve any changes in the status quo unless it generates the idea.

Develop a strategy for changing your software.

CHAPTER 12 Self-Development

KEY TERMS

education, p. 143
network, p. 146
self-development, p. 143
training, p. 143

Frequently, in the busy world of fire chiefs they forget to continue to develop their skills. Like everyone else in the organization, the chief needs to be as good as possible at doing the job. Clearly, skills, knowledge, and abilities are needed to perform at a high level. Some of these are practiced routinely, so performance levels are maintained. Other parts of the job are not practiced so frequently. In addition, the nature of the fire service is constantly evolving.

No one disputes that the world is changing rapidly. So is the job of administering and leading a fire department. There are issues today that just didn't exist 10 or 15 years ago. You certainly could think of some of them. If you accept the premise that the job will continue to evolve, then you must do the things necessary to keep yourself prepared.

More than ever the fire chief needs to understand the intricacies of the job and the things that make a chief successful. If you were to select a fire chief, you would want someone who is:

- Educated
- Ethical
- A communicator
- A planner
- A salesperson
- Flexible
- Creative
- Responsive
- Connected, or networked
- Political
- A team player
- Able to maintain balance and perspective
- Self-confident
- Enthusiastic

Of course, you would not want someone who stopped developing after accepting the job. As with every other position in the fire department, training and education needs to continue throughout the career. **Self-development** is critical to your future success. It is your investment in yourself to become more mature, so it is important to consider your education, training, and professional and personal development. Set your personal bar high so that you send a message of the importance of pursuing excellence. Elevate your game to set an example for the organization and allow everyone else to follow your lead.

self-development
Your investment in yourself to become more mature

◆ PLANNING

Regardless of whether you are the fire chief or are planning to be one some day, knowing what you need to do is essential; you need to have a plan. Your network and mentors can help you with this. In addition, resources are available through the International Association of Fire Chiefs, specifically its *Officers Development Handbook,* and the Center for Public Safety Excellence and its Commission on Professional Credentialing (CPC). Although nothing can guarantee your future as a fire chief, planning and preparation can help provide you the edge you need.

◆ EDUCATION AND TRAINING

First and foremost, continue your education and training until you retire. Understand the difference between training and education. **Education** expands your critical thinking skills and broadens your horizons. It can be defined as the systematic instruction designed to impart knowledge and develop skill, which is very important in the dynamic job of fire chief in which so much is unplanned and unscripted. **Training** is geared toward specific skills, changes in existing situations, or learning about emerging technology. It can be defined as a method of attaining a desired standard of efficiency or condition or behavior. As an example, at some point the fire chief has to learn the basics of computer use. Training is required to keep up with the changes and advances in this technology.

education
Systematic instruction designed to impart knowledge and develop skill

training
A method of attaining a desired standard of efficiency or condition or behavior

Education generally falls under the purview of colleges and universities (see Figure 12.1). You are encouraged to seek higher degrees than what you may already have. The degree may or not be specific to the emergency services. Even general education courses are helpful, as they make you a better-rounded individual better able to interact with others. The learning that will take place and the exercising of your mind are what are important. If additional degrees are not for you, then you can pursue other areas of formal education to continue your development. Many colleges and universities have very good outreach programs, and they may even offer continuing education credits should they be needed. Regardless of your motives, continual education adds to the diversity of your background and skills. Also consider college courses for the sake of learning even if they don't lead to a degree.

Besides formal education, there are numerous opportunities for enhancing your abilities through training. These can be fire- or emergency service–related programs or administrative/managerial/leadership types of courses. Opportunities are available through the National Fire Academy (NFA) and other fire service–related programs

FIGURE 12.1 ◆ College remains important in the development of fire chiefs.

FIGURE 12.2 ◆ The National Emergency Training Center (National Fire Academy) offers great opportunities for development (Courtesy FEMA/Jocelyn Augustino).

(see Figure 12.2). The NFA offers an outstanding Executive Fire Officer Program (EFOP). The EFOP covers many areas of importance to a fire chief and is a 4-year commitment involving 2 weeks of resident instruction each year and an applied research project each year. Many of the NFA programs, particularly those held on campus in Emmitsburg, Maryland, provide interaction with others, both classmates and instructors, which is important to personal development. The NFA also has outreach programs that are delivered locally and has developed a model curriculum for higher education. All these programs are professionally presented and beneficial to the fire chief.

The University of Maryland, through its Fire and Rescue Institute, offers the National Staff and Command School—a 1-week program, held each year—to help develop your skills. The program's condensed nature limits the time away from the job while providing excellent information and networking opportunities. Similar programs are offered by different institutions in many parts of the country. Information on such programs is often available from the state fire training agency.

Workshops and seminars are very beneficial not only for staying current with changes in the industry but also for helping develop a wider network. Topics can include the latest and greatest in the fire service or some of the basics that every fire chief needs such as budgeting, personnel management, legal issues, and labor relations. Regardless of the topic, the chance to "polish the apple" from time to time is good for the individual and organization.

◆ NETWORKING AND INFORMATION EXCHANGE

network
A chain of interconnected people

Part of your development is meeting new people and networking, both inside and outside the profession. A **network** is a chain of interconnected people. These can include professional groups, chiefs' associations and other organizations, as well as informal relationships that are good for you. There is much to be learned from others, and there are important contacts to be made. There are associations at all levels—local, state and national, and even international, including the International Association of Fire Chiefs, state chiefs' associations, and county groups (see Figure 12.3). Because all levels of government affect the delivery of your service, participation at different levels will offer unique benefits. There are probably more organizations than you will have time for, so consider the value of each. Whichever ones you choose, don't just join but participate, because there is much value in active association memberships. Besides gaining information, you get to meet professionals from similar departments with similar challenges. You may also get the chance to help influence policy and direction, which can have a direct impact on the services you provide.

Do not underestimate the value of conferences, workshops, and seminars. These can provide some education but may more closely align with training opportunities. These will allow you to stay current with the state of the art of your profession with respect to apparatus and equipment, management, administration, and government mandates. They can also keep you current on standards and other voluntary policies

FIGURE 12.3 ◆ The International Association of Fire Chiefs provides many services to fire chiefs and chief officers.

Figure 12.4 ◆ Attendance at conferences with vendor shows offers unique opportunities to see the latest innovations in the emergency services field.

that may affect your operation. Conferences such as Fire/Rescue International from the IAFC, the Fire Department Instructors conference sponsored by *Fire Engineering* magazine and PennWell, and Firehouse Expo are examples of national conferences that offer many learning opportunities. Usually there are also state and regional conferences along with specialty conferences (related to EMS, hazardous materials, terrorism, wildland fire, volunteer firefighters, etc.) that can enhance your self-development. Many of these conferences have a vendor show component where you can see what is being offered by the various manufacturers in the industry. If you are in the market for apparatus and equipment, this is one of the easiest ways to comparison shop (see Figure 12.4).

Don't always confine yourself to fire/emergency associations. Although the fire service is unique, it has similarities with other organizations that manage and administrate—both inside and outside government. Any opportunity that gives you a different perspective or provides insight or information to continue your development is good. You are encouraged to explore different options to determine what will be best for your development.

Consider local networking opportunities in your community. There are many great service clubs such as Optimist International, Rotary, Lions Club, and National Exchange Club. These provide opportunities for you to interact with your community, mingle with other professionals with different occupations, give back to your community, and gain additional skills, either directly or indirectly. The social contacts that you make are also beneficial.

◆ READING TO STAY CURRENT

You need to stay current in your profession, and reading is usually the simplest and most convenient way. Read whatever you can, including professional trade journals, textbooks (like this one), and publications about related issues outside your vocation. Not everything you read needs to have *fire, rescue,* or *emergency* in the title. The rapidly changing world of the emergency services makes journals and the Internet invaluable for staying current with issues. Government, laws, codes, and standards are constantly changing as well. Don't make a mistake because of ignorance resulting from failure to stay current. There is also value in continually exercising your mind.

◆ CONTINUAL CHANGE

One of the biggest challenges that fire chiefs face is staying current with changing laws, rules, and standards. For example, the NFPA publishes volumes of standards and recommended practices. Some have the force of law through adoption, whereas others establish best practices to be followed. You are expected to have a general understanding of applicable standards, especially if something goes wrong. In addition you must comply with local, state, and federal mandates, such as OSHA standards, which makes your job even more complex. All these are subject to change, so you need a system to stay as current as possible. Further, legal issues such as potential liability and labor relations issues are part of the information base that is critical to successfully administering a fire department. You cannot isolate yourself and need to pay attention to the world outside your organization through tools such as the aforementioned conferences, workshops, seminars, and journals, as well as the Internet. Your expanding network of contacts within the profession will be very helpful.

◆ MENTORING

Having a mentor and being a mentor are two good ways to continue with your self-development. First, you need to seek out someone who will help you, much like a coach. This person must be knowledgeable and respected and be able to give you unbiased advice and constructive criticism. You will be able to learn from his or her experiences and also bounce new ideas off him or her with the assurance that you will get useful feedback.

On the flip side, being a mentor can also help you. If you are to help others, you not only will rely on your experiences but will need to stay current yourself. The person you mentor will challenge you in many ways by asking questions and seeking your opinion. This will keep you current and a little on edge so you continue with your own development while helping someone else.

TEACHING, WRITING, AND SPEAKING

When you need to prepare to teach, write, or speak, you must learn the subject very well so that you do the best job you can. Learning as much about a subject will always be beneficial to your self-development. Besides producing the obvious benefit of learning a topic, it will exercise your mind and keep you current. Anything you can do to enhance your thinking abilities will make you a better chief. You may also get feedback from those you teach or who hear you speak or read your writings. This is very healthy.

LEARNING FROM OTHERS

What can you learn from others? A great deal, obviously. But don't expect everyone to come to you. Further, some of the lessons you learn will come not from discussions but from observation. Seek out those whom you respect. In addition to talking with them, observe their actions and behaviors. You will pick up many tips that can help your career and performance. Do not limit yourself to only those in the fire profession. You can learn much from others who are successful in government and business. Never miss these opportunities to expand your horizons. Following is a list of traits exhibited by successful fire chiefs. Although they are not supported by science or study, they may provide some insight as to things you can do to "polish" your apple. You may benefit by discussing these traits with others or adding to the list those you have observed in people you perceive to be successful at their job. They are most likely successful because they possess many of these traits.

Characteristic, Practices, Traits and/or Habits of Successful Fire Chiefs

1. Is a fire buff; has passion for the fire service; is a zealot
2. Has tremendous pride in the fire service
3. Is a politician
4. Has a respect for history
5. Has experience with fire—has been there, been in tough situations, done dumb things
6. Is genuine—does not try to be someone he or she is not
7. Does not brag about self but about the organization or the fire service
8. Likes challenges
9. Is not intimidated or in awe of anyone
10. Is caring, compassionate, and empathetic
11. Has good communication skills and is a good communicator
12. Is capable of making decisions; is able to pull the trigger
13. Has good timing in actions and decision making
14. Supports training
15. Has a fear of failure (This does not mean a fear of trying something new. It means the willingness and effort needed to attain success. It means that failure is not an option.)
16. Is concerned about reputation not only of self but, more important, that of organization and fire service
17. Commits time to the job; has tremendous dedication
18. Desires to continually improve; is able to take constructive criticism
19. Actively participates in various associations

20. Understands and respects labor relations
21. Is a big-picture thinker; is not concerned with technical detail or minutiae but sweats the small stuff
22. Change agent and accepts meaningful change, but not for change's sake
23. Takes advantage of one-time windows of opportunity and knows when to take advantage
24. Is willing to help peers and subordinates
25. Is involved in the community
26. Has fun, smiles, laughs; is able to poke fun at self
27. Has a positive attitude
28. Has a professional appearance
29. Follows through
30. Is trustworthy and honorable; keeps word; walks the talk
31. Is passionate

◆ SUMMARY

One of the easiest things to do is ignore your own development. Departments get busy, and other people and issues need your attention. But you cannot do your best unless you are best prepared. The good news is that you have demonstrated the knowledge, skills, and ability for the job. You just need to keep the "apple polished." Part of the responsibility of being a fire chief is to know your business. Knowing what it was in the past is not good enough. You need to be current and also prepared for the future, whatever it may bring.

ACTIVITIES

1. Identify three conferences related to your profession. Determine their applicability to your job and other factors that would affect your ability to attend, such as time and cost. Develop a plan to attend a conference.
2. Review three trade magazines and determine their benefit to your self-development.
3. Contact the National Fire Academy and determine the course offerings applicable to your self-development.
4. Identify local service clubs and consider attending their meetings. List the value of each to you personally.

Answers to Case Study Questions

The following "answers" to the Case Study Questions are sample solutions only. These are not an automatic fix as each situation is unique and it is not easy to classify and repair without knowing all the factors. Discuss the scenarios with your class or with a group to find an answer that is appropriate to your area.

CHAPTER 2

Obviously, the simple answer is to stay out of the race. Unfortunately, it often is not that easy. Depending on the politics of your community, the issue can get complicated. You may be pressured by others, not just the candidates. You may be linked to the firefighters. You may even have a personal opinion and a desire to support one of the candidates. One year out from the election, many issues may not be generated, but as the election gets nearer, and if the contest appears to be close, there may be more effort to drag you or the fire department into the campaigning either directly or through public safety issues, for example, fire service.

Your personal relationship with both candidates will be important. If you have a good relationship, you can be much more candid with your views, and they are more likely to respect your position, whatever it may eventually be. Your relationship may be such that you support one of the other candidates. (What would happen if your spouse was running against the mayor? Wouldn't you need to be public with your support?) A good relationship may allow you to support one of the candidates.

In some circumstances you may need to support a candidate. Your job may be a political appointment. The mayor, who appointed you and is now running, may have a hand-picked successor who could become mayor. The new mayor will then get to pick the fire chief! In some communities, you may be expected to let your choice be known. These are the types of situations that can be handled a little more easily if you have a good foundation in the community and you have earned the respect of many.

Avoid involvement in this situation as best you can. You may not know the relationships of all the players, and you could inadvertently offend someone. You may also affect your position. The department may also come under scrutiny, especially since the firefighters have chosen to support a candidate. In the political arena, most people do not distinguish among firefighters, the union, or the department. You may all be lumped together, and the decisions that one group makes will affect the others.

Of course, the situation can be even more complicated if you are not on the best of terms with the mayor or police chief. In that case you may be tempted to support one or the other, but you must resist the temptation as best you can. You never know who is in their network and who may be one of their supporters. They also have the potential to return to local politics at some future point. Make the right decision to protect your department and yourself now and in the future.

Finally, remember your own network. If you have done your job, you should know many of the players in your community. Talk to them. They can provide you with advice. They can suggest ways for you to handle the situation and also may offer you "protection" from any fallout. Don't forget your boss. Stay in touch and use his or her knowledge and experience.

On the surface, this seems like an easy issue. Once in a while, it is. Often in the political arena, however, things are not always as they seem and not as straightforward as they are in your department. Think things through. Do your homework and make a sound decision considering all the possibilities. You should end up okay if you don't make any hasty decisions.

CHAPTER 3

In many ways, appearance issues have not changed that much over the past 30 to 40 years. The length of hair, earrings, and other changes have challenged fire chiefs to maintain decorum and yet allow for freedom of expression. There must be a balance between what is needed to maintain the discipline and uniformity of the department and the individual's rights. Contrary to what some may believe, organizations do have the right to establish grooming standards for the good of the organization. You are cautioned not to have a knee-jerk reaction and to make any decisions with proper consideration and due diligence.

Resist the temptation to establish a policy that has not been adequately researched, and do your best to suppress your personal opinions. Whether or not you like tattoos or earrings should not affect the final decision. The policy must be based on what is good for the community and the department. It also must be legal. This does not mean that you cannot establish a reasonable policy, but you must do so using logic and reason.

You can approach this situation in one of two ways: do your own research or ask someone else to do it. Learning a little on your own may give you some guidance before you allow someone else to provide direction. You can learn of your rights and limitations by searching the Internet or literature. You also can contact your network—find out what your neighboring departments are doing, contact your peers and friends outside your region, and ask your professional associations for help. Most likely you are not in any rush, so take the necessary time to study the issue and potential options.

Others who are better prepared to address this issue than you may be are your human resource director or labor attorney. If you have a little background on the issues, you will be better able to discuss what you want and the best way to get there. As you do your research, remember to consult with both. They will help you with your research so that you do not inadvertently violate a law or statute. Appearance issues do not necessarily fall into a protected class unless they are related to religion, race, or other identified protected group. You need to know the limitations from the experts. Human resource personnel and city attorneys also may have dealt with similar circumstances and properly handled them in other departments or agencies. You will also want to inform your boss, as this type of issue can get messy, depending on the people involved.

If you have a labor union, this would be a good issue to partner on. The union may be more sympathetic to your position than you might expect and have the best interest of the department in mind. Union personnel are interested in uniformity and discipline. They are out in public with the firefighters more often than you and can be affected more frequently by the appearance of your members. They also have access to their own resources, which can be of help. In the end, if you work together, you will present a united front that will gain more support of the final policy.

Appearance issues can be tricky. They require good research and common sense. Others are available to help you. Remember that you can provide strong guidance in the development of the final policy so that you end up with something that you are comfortable with. Others will influence you and may try to exert pressure in support of their personal opinions. Be reasonable and fair, follow the law, and consider what is best for the community and the department.

CHAPTER 4

This is a clear example of an instance in which the fire chief is a middle manager, not the chief executive. Further, it represents a circumstance in which the chief must exhibit skills in following as well as in leading. As is the case with most negotiations, once an agreement is reached, the negotiating team moves on, and the chief must deal with the fallout. Although most people understand the nuances of negotiations, there can be hard feelings after a settlement is reached. The formal position of fire chief is clearly identified as being on the management team, even though the fire chief may be viewed as somewhat neutral. Positions taken and decisions made during negotiations can affect future relationships between the chief and union. The general state of morale can also be affected.

You need to clearly understand your role in negotiations, and that role must also be understood by others in the session, especially the union team. With that said, you need to work on your "poker face." Regardless of your personal beliefs, you need to maintain your position and not publicly react when issues are introduced and discussed. There will be times when you agree with management and times when you agree with labor. The best advice is not to allow these opinions to surface during the sessions (and not at all if possible). You will be asked to support management, even when you don't agree. If you display your displeasure, everyone will notice your reaction.

As in so many areas discussed in this text, strong relationships can help you through this process. If you continually work on your employee relations and those with the labor union, you will have a better chance of resuming normal operations after the negotiations are complete. You can approach the sessions as being part of the "business" and something that must be done without creating hard feelings or long-lasting damage. This is not easy, as feelings can be hurt and some things can be taken personally. Strong relationships allow for candid discussions and trust. You can be open and honest without fear of negative fallout.

Relationships are just as important with management. If you have good interactions with your boss, you can be honest and open without appearing to be on the "other side." You are bridging the divide between your boss and your position on the management team, and your role as leader of the fire department. If you enjoy a good reputation as a leader, and the reputation of the department is positive, you will have a stronger voice, and your opinions will be more respected. You are then in a position to express to your boss your thoughts regarding the progress of the negotiations. If you do not have such a reputation, you are going to be stuck, and you will assume your role and not do anything that could be deemed controversial. You should do what you can to maintain relationships with your employees without compromising your management position, as you will need to continue running your department at the conclusion of the negotiations.

Understanding the position of both sides will help. If you can, try to find out the goals of both sides in the negotiations. The municipality will have specific issues, mainly cost control. The union will obviously want enhanced benefits and pay but may also have other issues that are important to them. If you know what they are, you can help push the negotiations toward these issues.

Trust is very important. If you believe you can have frank discussions with both labor and management without breeching confidentiality, you are in a good position. If you do not have that trust, you will need to keep things to yourself and let the negotiations play out. You may not be confident that the management team can do the job. That is irrelevant. You will play with the cards you are dealt.

CHAPTER 5

There are two major players in this scenario. Both will be affected by the final outcome and must be accounted for in your plan. They are your employee group (firefighters) and the private ambulance company. Certainly, the policy makers in your community have a role, but their position will be influenced by the two main players.

Dealing with the firefighters will most likely not be that difficult, because they often see the value of the proposed change. They understand what is happening and what the benefits will be. Nevertheless, there likely will be issues that need to be resolved. For example, if new revenue is being generated for the department or community, who will reap the benefits? Will the money be placed in the general fund for the overall benefit of the community, or will it be returned to the fire department? Obviously the firefighters will expect the revenue to increase the department budget and increase staffing. This is a question that must be answered early in the process.

What will the proposed change mean to department operations? You will need to evaluate out-of-service time. Will it be increased because your units will spend more time at the hospital (completing paperwork, performing a drug box exchange, cleaning/sanitizing your apparatus and equipment, etc.)? If so, then you need to evaluate the benefits and justify this change in operation. Your personnel will need more training, especially in the part of the job they usually least like—paperwork! Information gathering will be very important to maximize your collections. You will also come under more scrutiny by those who pay the bills, especially the insurance companies. You will need to justify the services that were delivered and make sure protocols were followed.

The more challenging part of the change will be dealing with the private company. Most likely it is making a profit from the business, so it will be less likely to want to give it up. The company will compete to keep its business and will use almost any tactic to keep what it has. You will need the support of your boss and policy makers to move forward. You will need sound information and facts. Do not exaggerate your projections. It is best to be conservative on income estimates and to overestimate projected expenses. You will not want anyone to be able to question what you are presenting.

Know the players. Who on your policy-making board will be sympathetic to the private sector? Some politicians do not want government competing with the private sector. They believe government should do only what the privates can't. Understand this view and be prepared. Others will look to the potential revenue source for the community, especially if revenues are not growing. They may want the funding for their projects. Others may not want to charge for service, because they believe that taxes already pay for the service and so citizens should not be charged. You will need to do your homework so you are well versed and prepared. Finally, be aware of any political connections between the private company and your elected officials. Those personal bonds may not be well known and will present a challenge. Often, logic and facts do not change personal relationships. Understanding this dynamic will help with your plan.

Remember that logic and common sense do not always translate into change. In government and politics, things can be complex, and what may appear to you to be a good decision, may not be so attractive to others. You need to learn as much as you can about the players in addition to the nuts and bolts of your proposal. Change requires energy and effort. Good communications and trust are essential. If you are not willing to invest the time, don't expect good results.

CHAPTER 6

In today's world, cooperation between neighbors is essential to providing good service. Outside the largest departments, there is a need for mutual aid, as staffing is generally inadequate when many calls occur simultaneously or there is a large incident beyond the capabilities of a fire department. As in so many other circumstances, what has happened in the past will determine the outcome of the incident.

You need to get a feel for the severity of the problem directly from your shift commander, then use your judgment to decide whether to pursue this matter. If you decide to do so, obtain written statements from those who have particular knowledge of the incident. This allows you more time and information to evaluate the specifics of the incident. Keep in mind the need to be supportive of your personnel. They need to know that you will go to bat for them when the facts are on their side. The written statements will help establish those facts.

It is hoped that there is a good working relationship between chiefs. If this is the case, then resolving the issue becomes easier. Call the other chief to let him or her know what has occurred. Offer to meet with the chief and your shift commander. A strong statement by your personnel will demonstrate the need for action to correct the problem. Your side of the event will be explained with the written statements from your personnel and a meeting with the neighboring chief. You can expect the neighboring chief to do an investigation within his or her department.

Based on the findings of all parties, it may be necessary to have all affected personnel meet to express their opinions. Your goal is to resolve the problem and move on. These are the types of incidents that can leave hard feelings for a long time unless they are addressed so that the interdepartmental working relationships can be strengthened at every level of both organizations.

Perhaps the parties in the other department will admit to their indiscretions and offer an apology. If this is the case, be gracious and accept it. Do not let the bad feelings linger, and make sure your personnel are also gracious. If this is the resolution, then accept it and move on. Do not bring it up again.

If your personnel were overly sensitive, and it should have been a nonissue, address that problem. You may need to apologize, as might your personnel. Again, remember that future working relationships must be good, and you will need to work together in the future. Accept the outcome and move on.

This is a good opportunity to reinforce the need for good cooperation with all your members. One of your members may do something similar. Not all your members will like the added responsibilities of mutual aid, and they may not like all your neighboring departments. Remind everyone that they are expected to behave properly and professionally all the time. Rudeness is not an option. Problems should not occur on the scene. The emergency and patient, if there is one, take precedence, and differences must be resolved after the incident is handled.

Issues such as these may appear minor on the surface, and they may turn out to be relatively minor in the bigger picture. Regardless, if not investigated and acted on, they can affect future events. The rumor mill can spin this into a very large issue and affect relationships for a very long time. Take such incidents seriously. Defend your personnel as much as possible. Do what you can to clear the air. Shake hands and forgive indiscretions and set the stage for improved partnerships that will improve response to emergencies. In today's world, interdepartmental cooperation is essential, and your actions will set the tone.

CHAPTER 7

This is the type of incident that will affect your day-to-day operations but should not usually make it to your desk. There should be a system in place that allows supervisors to handle this issue without direction. They should be communicating only the nature of the event and how it was handled. Unfortunately, that is not always the case, and this is the type of daily interruption that will occur. Also remember that one of the most basic requirements for the department is to have enough people on shift. You are subject to questioning, so you have the ultimate responsibility.

When this matter is brought to your attention, you need to return it to the supervisor who should be handling this issue. Do not allow this to be "delegated" back to you so it becomes your problem. You need to make sure you don't start doing someone else's job. Once you start it is difficult to stop, and you may find more and more issues finding their way to your in-basket. Your initial action should be to provide simple instruction to the supervisor to follow procedure and handle the problem and report to you the outcome.

If this is a relatively new issue and the supervisor has no experience, it would be beneficial to provide some coaching as to the best way to address this issue. A little investment in time up front may save you much more later on. You want the supervisor to do what is right and in accordance with policy. A mistake will create more work for you later. Also, proper coaching will help when the next event happens. Just be careful not to open your door so wide that you become a crutch for the supervisor.

If this is the first time this happened, there are some things you can do (besides counting your blessings) to prevent future occurrences. If you do not have a policy on shift trades and responsibilities, you will need to draft one. It is a good chance to work with the union or other employee group to get consensus and buy-in so that everyone is in agreement on the best way to address this issue in the future. The policy must be very clear and communicated to the entire organization. One thing that is a must in the policy is a written form to be completed and signed by the parties to establish who is responsible to show up for the shift. You may wish to review it during a meeting and have the department members sign that they are aware of the policy. Don't forget to include this in your new-member orientation. You do not want anyone to claim that they were unaware.

If there are similar circumstances in which supervisors are not accepting their responsibilities, consider training. If there is a common theme, it might mean they just don't know what their job is. Often, employees are promoted and it is assumed they know what to do. In the vast majority of cases, new officers or supervisors want to do the best they can. You need to make sure you have given them the training and tools to do a good job. Review your practices and make sure your message is being delivered.

This is not a major issue in the larger picture of running a fire department. It is not a large challenge or a time consumer. Rather, it is another event during the day-to-day operation of your organization that requires your attention. It takes you away from other parts of your job and adds to your workload an item that is best handled by someone else. Fix the immediate problem—most likely to hold the person who accepted the trade responsible—and coach your supervisors so that future issues are theirs, not yours.

CHAPTER 8

In the overall scope of things, the handling of this incident caused no real problem. Most likely there was no threat, and no harm resulted. The issue is that an opportunity to add value was missed. The job of your department is to deliver the type of service that will be remembered and to develop loyalty within your community for your department, not just to provide minimum service.

Do you have a policy on how these incidents are to be handled? If not, you can expect your personnel to make decisions based on their values and the training and education they received. They may have learned what to do from a previous officer. If you leave choices up to individuals, you are likely to get a variety of responses. If you desire predictable and consistent response, establish a policy and procedure.

If the situation was not handled as you desire, or if there is a policy in place, you need to have a discussion with the shift commander to change the behavior. If this is a first offense, simple counseling may fix the situation. If there is a pattern of bad judgment, "rehabilitation" may be necessary. Regardless of the personal opinion of the individual, you need to change the thinking for future incidents.

Training for all department personnel may be in order to cover the topic of customer service. Much is said about it but little is invested. There is an assumption that everyone is good at it and willing to participate, but communication and training are necessary. Let all your members know what is expected and train them to those expectations. Most people will want to do a good job, but they need help in determining what exactly top-of-the-line service is. Often, customer service is a learned behavior, and individuals learn from previous supervisors. If someone was fortunate enough to have a leader who understood and communicated the virtues of customer service, then that person is likely to carry on the tradition. If the boss was lacking in this area, the next generation is also likely to be lacking.

With respect to the actual incident that started this discussion, it probably is a good idea to follow up and really put the department's best foot forward. You do not need to acknowledge any faults in your initial response but can bend over backward to demonstrate your concern as a department for the well-being of the caller. Often, it is the compassion demonstrated by the firefighters that is most remembered. Allow your personnel to respond and create a new, loyal customer, someone who will be a lifelong supporter.

Customer service is sometimes considered a buzz word, but it shouldn't be. It means delivering the best you can each time you respond. Added value builds support and loyalty to your organization. Not everyone you hire has the natural skills to be great in this arena. Create a culture of good service by educating and training your firefighters so they know and understand how to enhance what they do. The public expects the fire department to be good. When it exceeds expectations, there is no downside, only positive results.

CHAPTER 9

Finding the balance between getting things done by cutting through governmental red tape and following procedures can be a challenge for an emergency service business that is expected to respond at a moment's notice. Regardless, all members of government are expected to follow the rules. Shortcuts in purchasing and other areas of finance can get you in trouble. Although it is difficult to criticize members who are passionate about the job and effective in facilitating needed acquisitions, you need to draw the line.

The first step is instituting damage control with the purchasing director. Like it or not, those in charge of acquisitions have a great deal of control over your ability to get the resources you need. Although you may be able to build up enough political support to overcome poor relationships, this is not the best way to handle the circumstances. You are best served by developing a good sense of cooperation and a relationship of openness and trust. This will allow for occasional deviations from policy to be handled appropriately and without fanfare. You need to apologize for this indiscretion and indicate that you will do the best you can to avoid a repeat. You also need to ask if there is something that can be done to fix this situation and purchase. You need to be sincere and make good on your promise to do better.

This is a case in which relationships are very important. A good relationship with those who contribute to your operation will allow you to resolve the problem at the appropriate level. A poor relationship may cause this problem to be brought to the attention of your boss. Frequent instances of violations of city policies can get you in trouble. It is likely that you will make an error from time to time. Therefore, your ability to work it out with the affected parties is important.

Don't forget about your employee. You need to determine the cause of the problem. Was it laziness or a lack of understanding of the policy? Is this something that has happened frequently in the past but has not been caught? Whatever the reason for the problem, it needs to be fixed. If it was an honest mistake, make note of it and move on. If there is a pattern, you may need to consider a reprimand. The employee may also need to apologize to the purchasing director. Consider the history of the employee. Good performers with a solid history should be given the benefit of the doubt. If they are overachievers, they are of value to the organization and need to remain motivated. This is part of that fine line—keeping your personnel, especially the ones extending themselves, motivated is very important. Keeping them operating within the policies of the community is equally important. Good counseling and mentoring are probably a better option than a reprimand unless there is flagrant disregard for the policy. In this case, there did not appear to be an emergency, and it may have been a simple case of taking a shortcut. Don't let this be repeated.

Sometimes, training of personnel in the workings of the city system is neglected. There are policies in all areas. If you are going to assign staff to functions that involve working with others and their procedures, make sure they know the procedures. Also, make sure you encourage your staff to build relationships with those with whom they will be working. If cooperation exists on their level, you may never even hear of the problem, as it will be resolved between the parties directly involved.

As the fire chief you need to realize that you are ultimately responsible for the actions of your personnel. When issues arise outside your organization, you will be scrutinized even though it may have been one of your employees who caused a problem, and you had no knowledge of it. The buck stops with you. You can minimize the number of incidents if you train your personnel, communicate your wishes, make sure everyone involved knows the policies and procedures, and encourage the building of personal relationships among all involved.

CHAPTER 10

Like so many decisions that you must make, the initial analysis appears simple and straightforward, but you usually have time to investigate and possibly make a better decision than your initial reaction produced. On the surface, the quick answer is to return the wine and say "thanks, but no thanks" as tactfully as you can. In many cases that will be the right answer and may even prove to be so after an evaluation. As with many ethical questions, there is no standard answer; no one can one say there is only one correct answer.

One way to start discussion of this issue is to ask whether the department has ever accepted a thank-you gift after a call. In many cases the answer is yes. Some departments have received home-baked goods attached to notes from grateful citizens. Others may have received a cash donation to their favorite charity or to the general fund. If that has happened, the question then becomes whether there is a value limit to the gift.

Now, look at the situation from the store owner's point of view. He is just as grateful for the work of your members as the person who sends in a thank-you note with a "little" present attached. He has no ill intentions as to expectations. He is very sincere and appreciative, and it may be part of his recovery process to express thanks in such a way that helps him move on. Regardless of the reason, a property owner wants to show appreciation in the only way he knows how. Is this wrong?

Departments must be prepared for ethical issues. The first thing that needs to be done is to draft a policy. There are many great examples. They vary from community to community and department to department. You need to have a policy that is acceptable in your community. It is probably best to have a citywide policy, as opposed to one specific only to the fire department, to provide consistency within the municipal government. If your community does not have a policy, you will need to draft one for your own benefit. Your membership needs to know the rules surrounding ethical issues. As with all policies, you must make sure that your members know and understand them. You may wish to consider a training program or some other means to ensure that your members are aware of the policy and its contents.

Now, let's return to the wine issue. It now becomes easier to act because you have a policy. There may be extenuating circumstances. The note may have been anonymous. The owner may become offended at any offer to return the gift. This does not give you permission to violate policy, but it may require more effort on your part. We continue to emphasize that there is no need to rush, so investigate this issue from all angles.

Don't be afraid to discuss this with your boss. He or she is a great resource and can be of big help. It also gets the issue in the open. Openness with ethical issues is a great way to come to resolution. Although you may enjoy a glass of wine, whether or not you keep this gift for the department will not affect your operation. It is the way you handle it that will make a difference. Others in your community may offer advice. Consider the cultural or ethnic issues that may be part of this problem. You need to make decisions based on the good of the community and the department.

Now that you have the issue out in the open, do you have multiple solution options? You can return the wine, keep it for the department, or give it away. Returning it may cause more problems than it solves. You and your resources will have to determine whether this is the case. If you keep it, let it be known what you did to all—the department members and your boss. This is no time to keep secrets that can blow up later. If you keep the wine, there must be an equitable method of distribution. Think of it as "cookies from grandma." What would you do with them? Another option is to regift the wine. Is there an opportunity to help a charity or someone else? Perhaps there is a way to auction the wine that would benefit a needy cause. It appears that you may have some choices.

Make the best choice you can based on the facts, your policies, and the advice of your trusted advisors. Announce your decision openly, and then do what you said you would do. Whatever you do, do not keep any of this a secret. Keeping everything in the open is usually the best policy regarding these types of issues.

CHAPTER 11

Technology is changing very rapidly, so the fire chief is challenged to best utilize technology and adapt as the changes occur. The chief cannot change every time there is a new product—there could be monthly changes—but must stay on top of the industry and consider when a new or improved product is necessary. Software issues present a unique challenge.

The fire chief has many ways to stay current on trends in software, specifically, a records management system: through trade journals, trade shows, other departments, sales representatives, and other members of the department. As the world of computers becomes more complex few fire chiefs (and probably few firefighters) can keep up and know all the aspects of a network computer system. Because records management must fit into the current computer system, it must be evaluated by specially trained personnel. These experts are often not part of the fire department but are part of either an IT department or a consulting business. The fire chief will need to work through these people when considering computer usage.

It is not often that independent departments such as IT move as fast on fire department issues as the chief would like. IT personnel have other job responsibilities and other department heads who also want their service. Though they most likely were hired to support the IT needs of the entire municipal organization, they have choices on how to utilize their limited time and resources.

The fire chief may not have a direct working relationship with IT. Someone else in the fire department usually has the daily responsibility for IT functions and interaction with the IT department. The role of the chief is to facilitate needed changes and offer the necessary support to move the project forward. Even though the chief desires the change, this does not guarantee that there will be immediate movement, as IT people have the chief at a distinct disadvantage. Consequently, changes in this area are dependent as much on the chief's ability to exert influence as they are about product knowledge.

A strong relationship is necessary between the fire department and IT department. Like all other relationships, this one is built over time, not only when service is needed. In addition to the fire chief others in the fire department must help establish a strong foundation of mutual respect and trust. Remember, that often, nothing happens in this discipline without the recommendation of the city's experts. Regardless of what you or your members think, changes will not occur without the blessing of these experts.

If you have a good working relationship, initiate a meeting. Include IT and your personnel right from the start. Explain your position and what you would like to see happen, provide direction, and turn the experts loose. Establish a budget and a deadline and provide any resources and information on the product or products you think would be good. More than likely, other fire department members will be driving this issue, and you will be the broker for this deal, facilitating the change. Do not think you can order this to happen. You are making a request and must rely on your ability to influence the final decision making. You may be the final decision maker, but you will not get a chance to decide until others make their recommendation. Once the recommendation is made, it is best to approve it. Of course, you have a fiduciary responsibility to know enough about the issue to make a good choice.

You will rarely win an argument, debate, or discussion with IT personnel regarding computers and software. To get things accomplished, you must work through others. There is little need for technical knowledge. The important skill here is the ability to build long-term relationships with those who provide support to the fire department. This is a continual process because of changing technology, changing personnel, changing relationships, and the like, and requires time and effort. This commitment will be rewarded with a high quality of service from the specialists. Invest the time and energy to know the issues and the players involved.

Glossary

apparatus The "rolling stock," that is, vehicles such as engines, ladders, rescues, squads, and ambulances

background check A look into the history of the candidate relevant to the position being sought

beliefs Statements, principles, or doctrines that a person accepts as true

binding arbitration A process for settling a dispute between two parties in which both sides present their best last offer, and the assigned arbitrator hears the arguments and make a ruling

budget An estimate of the income and expenses of the community

communication The exchange of information by means of speaking, writing, listening, and even unspoken signs or behaviors

culture A particular set of attitudes that characterizes the fire department

customer service Work done for clients as a job, duty, or favor

delegate To give someone the authority to act on your behalf

discretionary Freedom to make decisions based on the circumstances.

distraction Anything that interferes with concentration or takes attention away from something else

document A formal piece of writing that provides information or that acts as a record of events

due process Formal proceedings carried out in accordance with the established rules

education Systematic instruction designed to impart knowledge and develop skill

effective Producing the desired results

efficient Able to function without waste

equipment The tools needed for the purpose, activity, or job at hand

ethics The study of moral standards and how they affect conduct

fact finding Presentation of issues as parties view them to an unrelated third party who then considers both sides of the dispute and makes a recommendation

fees Payments for some forms of professional service

finance The business of managing the monetary resources of the community

grievance A cause for a complaint

hiring criteria Standards used in the selection of an employee

International Association of Fire Fighters (IAFF) The major representative of firefighters in the United States and Canada

interruptions Breaks in activities that temporarily halt work

job description A written document identifying the skills, knowledge, and abilities needed for a position

labor agreement (contract) A formal agreement on wages, benefits, and working

leadership The ability to guide, direct, or influence people conditions (mostly those related to safety)

mandatory Required by ordinance or by the people

marketing The selling of a product or service

MBWA Managing by wandering around

media Television, newspapers, and radio collectively

mediation A process designed to resolve a dispute between two opposing sides

meetings Occasions where people gather to discuss something

morals Involving issues of right and wrong

motivate To create enthusiasm, interest, and commitment

network A chain of interconnected people

open-door policy Free and unrestricted access

paradigm Typical example

partisan politics Strong allegiance to a particular political party

performance The effectiveness of the way personnel do their job

physical ability tests A series of tests administered to determine a candidate's strength, endurance, and agility with respect to job performance

political relationships Good dealings with those influential in the political arena

politician A person formally engaged in politics

politics The art or science of government; the art or science concerned with guiding or influencing government policy; competition between competing interest groups or individuals for power and leadership; the total complex of relations among people living in society

prerequisites Conditions that must be met prior to application for a position

probationary period A defined period of time during which a person's behavior and ability is observed and tested

professionalism The skill, competence, or character expected

proficiency The competence of personnel at their job

purchasing Buying something using money

rumors Idle speculation or unverified reports

self-development Your investment in yourself to become more mature

strategy An overall plan

tactics The means to accomplishing the goals of a plan

taxes The amount of money levied by the government on its citizens and used to operate the government

technology The study, development, and application of devices, machines, and techniques

training A method of attaining a desired standard of efficiency or condition or behavior

trust Confidence that everyone will be treated fairly, truthfully, honestly, and with respect

values The accepted principles or standards of an individual or group

Index

D

E

I

J

K

L

U

V

W